Asterisk 1.6

Build feature-rich telephony systems with Asterisk

David Merel

Barrie Dempster

David Gomillion

BIRMINGHAM - MUMBAI

Asterisk 1.6

Copyright © 2009 Packt Publishing

All rights reserved. No part of this book may be reproduced, stored in a retrieval system, or transmitted in any form or by any means, without the prior written permission of the publisher, except in the case of brief quotations embedded in critical articles or reviews.

Every effort has been made in the preparation of this book to ensure the accuracy of the information presented. However, the information contained in this book is sold without warranty, either express or implied. Neither the authors, Packt Publishing, nor its dealers or distributors will be held liable for any damages caused or alleged to be caused directly or indirectly by this book.

Packt Publishing has endeavored to provide trademark information about all the companies and products mentioned in this book by the appropriate use of capitals. However, Packt Publishing cannot guarantee the accuracy of this information.

First published: September 2009

Production Reference: 1160909

Published by Packt Publishing Ltd.
32 Lincoln Road
Olton
Birmingham, B27 6PA, UK.

ISBN 978-1-847198-62-4

www.packtpub.com

Cover Image by Raghuram Ashok (raghuram.ashok@gmail.com)

Credits

Authors
David Merel
Barrie Dempster
David Gomillion

Reviewer
Justin Thomas Zimmer

Acquisition Editor
Louay Fatoohi

Development Editor
Swapna V. Verlekar

Technical Editors
Conrad Sardinha
Neha Patwari

Copy Editor
Sanchari Mukherjee

Indexer
Rekha Nair
Hemangini Bari

Editorial Team Leader
Gagandeep Singh

Project Team Leader
Priya Mukherji

Project Coordinator
Ashwin Shetty

Proofreader
Chris Smith

Graphics
Nilesh Mohite

Production Coordinator
Shantanu Zagade

Cover Work
Shantanu Zagade

About the Authors

David Merel is the founder and CEO of **Thinkbright LLC** a local/long distance telephone company as well as a cutting-edge Voice over IP carrier, providing businesses of all sizes with sophisticated and low cost VoIP solutions.

David started Thinkbright (www.thinkbright.net) in 2005 and continues to manage the company and its employees, all of whom are dedicated IT professionals. David acts as the company's chief architect, continually designing new technologies that have added significant revenues to the company's operations. During his many years at Thinkbright, David has worked with the latest Voice over IP technology, including all VoIP equipment from major manufacturers such as Cisco, Polycom, Grandstream, and Aastra. He also works with customers ranging from small businesses to Fortune 500 companies, and interacts with system integrators and IT consultants who call Thinkbright on a daily basis for assistance with all the VoIP solutions that Thinkbright offers. Thinkbright manages its own PBX system, providing customers with PBX features such as Auto Attendants, Waiting Rooms, and Ring Groups, or assists customers in managing their own PBX network while providing these customers with the service for incoming and outgoing calls.

David has many years of experience with Trixbox and Asterisk, and has installed countless custom configurations and deployments using those solutions. He also reviewed *Trixbox 2.6*, which is an excellent complimentary book to *Asterisk 1.6*.

David earned a Bachelor of Arts triple majoring in Philosophy, Politics, and Law from SUNY Binghamton. David holds a CCNA (a Cisco Certified Network Associate) certificate and is proficient in over 10 programming languages and databases, various operating systems, VoIP and related protocols, and other business applications.

> I would like to thank Samantha Brinn and Tony Shi who have helped in producing this book. Samantha Brinn, who assisted in much of the grammatical and style editing, and Tony Shi who conducted research on many of the Asterisk installation steps discussed in the book.

Barrie Dempster is currently employed as a Senior Security Consultant for NGS Software Ltd, a world-renowned security consultancy well known for its focus in enterprise-level application vulnerability research and database security. He has a background in Infrastructure and Information Security in a number of specialized environments such as financial services institutions, telecommunication companies, call centers, and other organizations across multiple continents. Barrie has experience in the integration of network infrastructure and telecommunication systems requiring high-caliber secure design, testing, and management. He has been involved in a variety of projects from the design and implementation of Internet banking systems to large-scale conferencing and telephony infrastructure, as well as penetration testing and other security assessments of business-critical infrastructure.

David Gomillion currently serves as Director of Information Technology for the Eye Center of North Florida. There he orchestrates all of the technological undertakings of this four-location medical practice, including computers, software (off-the-shelf and custom development), server systems, telephony, networking, as well as specialized diagnostic and treatment systems. David received a Bachelor's of Science in Computer Science from Brigham Young University in August, 2005. There he learned the theory behind his computer experience, and became a much more efficient programmer. David has worked actively in the Information Technology sector since his freshman year at BYU. He has been a Networking Assistant, an Assistant Network Administrator, a Supervisor of a large Network and Server Operations unit, a Network Administrator, and finally a Director of Information Technology. Through his increasing responsibilities, he has learned to prioritize needs and wants, and applies this ability to his Asterisk installations.

About the Reviewer

Justin Thomas Zimmer has worked in the contact center technology field for over 10 years. During this time, he has performed extensive software and computer telephony integrations using both PSTN and IP telephony. His current projects include system designs utilizing open source soft switches over more traditional proprietary hardware-based telephony, and the integration of these technologies into market-specific CRM products.

As the Technical Partner of Unicore Technologies out of Phoenix, AZ, Justin is developing hosted contact center solutions for the low-end market. Unicore's solutions present contact centers with low startup costs in a turbulent economy, and allow those centers to scale their business while maintaining a consistent and familiar user interface.

He has worked on countless software user manuals and instructional guides for both internal and customer usage. He has reviewed the book, *FreePBX* published by Packt Publishing.

He has also worked on *The Hopewell Blogs*—a science fiction adventure novel that will be released chapter by chapter online and available in print once the final chapter has been released.

> I'd like to thank the countless community contributors who have provided enough online documentation to make this book as accurate and helpful as possible. And I'd like to thank my wife Nicole for putting up with the extra hours spent reviewing this book, as well as my boys Micah, Caden, and daughter Keira for giving up some of their daddy-time for this project.

Table of Contents

Preface	**1**
Chapter 1: Introduction to Asterisk	**7**
What is Asterisk?	**7**
What's new in Asterisk 1.4?	**8**
What's new in Asterisk 1.6?	**9**
Asterisk is a PBX	9
Extension-to-Extension calls	9
Line trunking	10
Telco features	10
Advanced Call Distribution	11
Call Detail Records	11
Call recording	12
Call parking	12
Call barging	13
Asterisk is an IVR system	13
Asterisk is a call center system	13
Asterisk is a voicemail system	14
Asterisk is a Voice over IP (VoIP) system	14
Asterisk 1-2-3	16
Asterisk scalability	18
Asterisk does not run on Windows	19
Is Asterisk a good fit for me?	**19**
Trade-offs	19
Flexibility versus ease of use	19
Graphical versus configuration file management	20
Calculating total cost of ownership	21
Return on Investment	22
Summary	**23**

Table of Contents

Chapter 2: Making a Plan for Deployment — 25
The Public Switched Telephony Network (PSTN) — 25
Connection methods — 25
- Plain Old Telephone Service (POTS) line — 26
- Integrated Services Digital Network (ISDN) — 26
- T1 or E1 — 27
- Voice over IP connections — 28

Determining our needs — 29

Terminal equipment — 31
Types of terminal devices — 31
- Hard phones — 31
- Soft phones — 35
- Analog adapters — 36
- Another PBX — 37

Choosing a device — 38
- Features, features, and more features… — 38
- Determining true cost — 39
- Compatibility with Asterisk — 40
- Sound quality analysis — 40
- Usability issues — 41

Recording decisions — 41

How much hardware do I need? — 42
Choosing the extension length — 43
Preparing a test environment — 46
Summary — 46

Chapter 3: Installing Asterisk — 47
Preparing to install Asterisk — 47
Obtaining the source files — 48
Installing DAHDI — 49
Installing LibPRI (optional) — 49
Installing Asterisk — 50
Getting to know Asterisk — 54
Summary — 56

Chapter 4: Configuring Asterisk — 57
DAHDI interfaces — 58
system.conf — 58
- Lines — 59
- Terminals — 63

chan_dahdi.conf — 63
- Lines — 68
- Terminals — 68

SIP interfaces — 70
IAX interfaces — 74

Voicemail	77
Music on hold	80
Queues	81
Conference rooms	83
Summary	84
Chapter 5: Creating a Dialplan	**85**
Creating a context	85
Creating an extension	87
Creating outgoing extensions	92
Advanced Call Distribution	96
Call queues	96
Call parking	100
Direct Inward Dialing (DID)	101
Automated attendants	103
System services	106
Summary	109
Chapter 6: Quality Assurance	**111**
Call Detail Records	112
Flat-file CDR logging	112
Database CDR logging	113
Monitoring calls	116
Recording calls	118
Legal concerns	119
Summary	120
Chapter 7: Making Asterisk Easy to Manage	**121**
Trixbox	122
CentOS	122
Trixbox preparation and installation	122
What is FreePBX?	124
FreePBX preparation and installation	125
FreePBX System Status Dashboard	131
Tools	132
Setup	133
Trixbox maintenance section	135
Reports	136
Asterisk Recording Interface	137
Flash Operator Panel (FOP)	137
Flash operator configuration files	139
Web MeetMe	139
Setting up and accessing Web MeetMe through Trixbox	140
Flexibility when needed	143

A simple one-to-one PBX	**143**
Extensions	144
Trunks	145
Routes	146
Summary	**147**
Chapter 8: What is asterCRM?	**149**
Installing asterCRM	**150**
Automatic installation	151
Manual installation	151
Introducing asterCRM	**157**
Import	157
Statistic	158
Extension	158
Customer	159
Dialer	159
System	159
Survey	160
Summary	**160**
Chapter 9: Case Studies	**161**
Small office/home office	**161**
The scenario	161
The discussion	162
The configuration	162
system.conf	163
chan_dahdi.conf	163
musiconhold.conf	164
voicemail.conf	164
modules.conf	165
extensions.conf	165
Conclusions	166
Small business	**167**
The scenario	167
The discussion	167
The configuration	168
system.conf	168
chan_dahdi.conf	169
musiconhold.conf	169
agents.conf	169
queues.conf	170
sip.conf	171
meetme.conf	172
voicemail.conf	172
extensions.conf	173

Conclusions	178
Hosted PBX	**178**
The scenario	178
The discussion	179
The configuration	179
system.conf	179
chan_dahdi.conf	179
musiconhold.conf	180
sip.conf	180
voicemail.conf	181
extensions.conf	182
Conclusions	185
Summary	**185**
Chapter 10: Maintenance and Security	**187**
Backup and system maintenance	**187**
Backing up configurations	188
Backing up voice data	191
Backing up log files	191
Backup scripts	192
Time synchronization	195
Adding it all to cron	195
Back up Asterisk with FreePBX	196
Back up Asterisk with Trixbox	197
Rebuilding and restoring the Asterisk server	197
Disaster Recovery Plan (DRP)	198
Asterisk server security	**199**
Internal access control	199
Host security hardening for Asterisk	**201**
Integrity checker	202
Rootkit detection	202
Automated hardening	202
Role Based Access Control (RBAC)	203
Network security for Asterisk	**204**
Firewalling the Asterisk protocols	204
SIP (Session Initiation Protocol)	205
H.323	206
IAX	206
The Real-Time Transport Protocol (RTP)	206
Controlling administration of Asterisk	207
Asterisk scalability	**208**
Load balancing with DNS	209
Support channels for Asterisk	**210**
Mailing lists	211

Forums	211
Internet Relay Chat (IRC)	212
Web sites	212
Digium	212
Summary	**213**
Index	**215**

Preface

Asterisk is a powerful and flexible open source framework for building feature-rich telephony systems. As a Private Branch Exchange (PBX), which connects one or more telephones, and usually connects to one or more telephone lines, Asterisk offers very advanced features, including extension-to-extension calls, queues, line trunking, call distribution, call detail rerecords, and call recording. This book will show you how to build a telephony system for your home or business using this open source application. *Asterisk 1.6* takes you step-by-step through the process of installing and configuring Asterisk. It covers everything from establishing your deployment plan to creating a fully functional PBX solution. Through this book you will learn how to connect employees from all over the world as well as streamline your callers through Auto Attendants (IVR) and Ring Groups.

This book is all you need to understand and use Asterisk to build the telephony system that meets your need. You will learn how to use the many features that Asterisk provides you with. It presents example configurations for using Asterisk in three different scenarios—for small and home offices, small businesses, and hosted PBX.

Over the course of ten chapters, this book introduces you to topics as diverse as the Public Switched Telephony Network (PSTN), Voice over IP Connections (SIP / IAX), DAHDI, LibPRI, through to advanced call distribution, automated attendants, FreePBX, and asterCRM.

With an engaging style and excellent way of presenting information, this book makes a complicated subject very easy to understand.

What this book covers

Chapter 1: *Introduction to Asterisk* introduces you to Asterisk and goes over certain basics such as its capabilities and features, requirements, scalability, and cost of deployment. In addition, it covers the trade-offs of Asterisk, its benefits, and how to determine whether it can fit your needs.

Chapter 2: *Making a Plan for Deployment* goes over the planning of equipment needed such as phones and adapters, the phone service you will use to power your Asterisk server such as PSTN lines or a SIP service from a VSP. Other planning such as hardware requirements and conducting a sound quality analysis are covered as well.

Chapter 3: *Installing Asterisk* shows you how to install Asterisk using the source files. A step-by-step tutorial will take you through the entire process.

Chapter 4: *Configuring Asterisk* shows you how to connect your Asterisk server with either your phone service (through PSTN or SIP, among others) as well as how to deploy some basic PBX features such as queues, voicemail, and music on hold. Other advanced features such as faxing, video conferencing, and using conference rooms are also covered in this chapter.

Chapter 5: *Creating a Dialplan* focuses on creating your dialplan that determines how your calls are routed through the Asterisk server. You will learn how to create extensions, distribute calls in an orderly manner using queues, and present callers with a greeting using automated attendants (IVR).

Chapter 6: *Quality Assurance* tells us everything regarding monitoring calls, recording calls, and capturing detailed call logs. Here you learn how to install and use these features.

Chapter 7: *Making Asterisk Easy to Manage* shows you how to integrate third-party applications that make Asterisk easily manageable all through a web-based interface. The chapter will show you how to install FreePBX as well as give you an overview of its capabilities.

Chapter 8: *What is asterCRM?* tells us about a useful business application—a customer relationship management system (CRM) called asterCRM. Given its open source nature, Asterisk is compatible with many other business open-source applications. This chapter goes over installing and using this application, which can help streamline your business operations.

Chapter 9: *Case Studies* discusses several case studies to give you real-world examples of how one would deploy Asterisk. The examples will give a summary of the deployment as well as the asterisk configuration code used to carry it out.

Chapter 10: *Maintenance and Security* is an important chapter as it focuses on keeping your Asterisk system running smoothly as well as keeping it secure. The chapter covers security precautions, network deployment recommendations, as well as maintenance tips such as backups and preparing disaster recovery plans.

What you need for this book

Even though this book will provide you with step-by-step instructions, it is best if the reader has a basic understanding of Linux and its commands. For implementing Asterisk, you will need a PC with a P4 CPU or higher, 1 or 2 GB of RAM, and a hard drive of no less than 60 GB, 7200 RPM. Please note that it is possible to run Asterisk on lesser requirements; the configuration mentioned is simply a recommendation.

Who this book is for

This book is aimed at anyone who is interested in building a powerful telephony system using the free and open source application, Asterisk, without spending thousands of dollars buying a commercial and often less flexible system.

This book is suitable for the novice and those who are new to Asterisk and telephony. Telephony or Linux experience will be helpful, but not required.

Conventions

In this book, you will find a number of styles of text that distinguish between different kinds of information. Here are some examples of these styles, and an explanation of their meaning.

Code words in text are shown as follows: "Asterisk provides a number of defaults and we can configure additional ones in the /etc/asterisk/indications.conf file."

A block of code is set as follows:

```
[default]
mode=files
directory=/var/lib/asterisk/music-on-hold
random=yes
```

Any command-line input or output is written as follows:

```
#!/bin/bash
$ tar xjvf asterisk_backup.tar.bz2
$ cp -R etc/asterisk /etc/asterisk
```

Preface

New terms and **important words** are shown in bold. Words that you see on the screen, in menus or dialog boxes for example, appear in the text like this: "Once installed you will find the **Backup & Restore** module located under the **Tools** section of the GUI ".

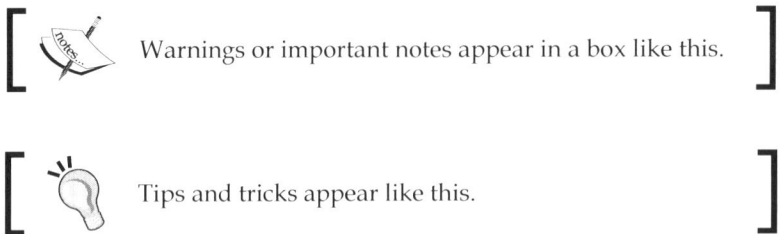

> Warnings or important notes appear in a box like this.

> Tips and tricks appear like this.

Reader feedback

Feedback from our readers is always welcome. Let us know what you think about this book—what you liked or may have disliked. Reader feedback is important for us to develop titles that you really get the most out of.

To send us general feedback, simply send an email to feedback@packtpub.com, and mention the book title via the subject of your message.

If there is a book that you need and would like to see us publish, please send us a note in the **SUGGEST A TITLE** form on www.packtpub.com or email suggest@packtpub.com.

If there is a topic that you have expertise in and you are interested in either writing or contributing to a book on, see our author guide on www.packtpub.com/authors.

Customer support

Now that you are the proud owner of a Packt book, we have a number of things to help you to get the most from your purchase.

Downloading the example code for the book

Visit http://www.packtpub.com/files/code/8624_Code.zip to directly download the example code.

The downloadable files contain instructions on how to use them.

Errata

Although we have taken every care to ensure the accuracy of our content, mistakes do happen. If you find a mistake in one of our books—maybe a mistake in the text or the code—we would be grateful if you would report this to us. By doing so, you can save other readers from frustration, and help us to improve subsequent versions of this book. If you find any errata, please report them by visiting http://www.packtpub.com/support, selecting your book, clicking on the **let us know** link, and entering the details of your errata. Once your errata are verified, your submission will be accepted and the errata added to any list of existing errata. Any existing errata can be viewed by selecting your title from http://www.packtpub.com/support.

Piracy

Piracy of copyright material on the Internet is an ongoing problem across all media. At Packt, we take the protection of our copyright and licenses very seriously. If you come across any illegal copies of our works, in any form, on the Internet, please provide us with the location address or web site name immediately so that we can pursue a remedy.

Please contact us at copyright@packtpub.com with a link to the suspected pirated material.

We appreciate your help in protecting our authors, and our ability to bring you valuable content.

Questions

You can contact us at questions@packtpub.com if you are having a problem with any aspect of the book, and we will do our best to address it.

1
Introduction to Asterisk

In this chapter, we will look at what Asterisk is and what it can do for us. As we explore features, we can make note of what features will help us to accomplish our goals.

What is Asterisk?

This is a fascinating question—what exactly is Asterisk? There are a number of answers, all of which are accurate.

First, Asterisk is a symbol which is denoted as *. The symbol represents a wildcard in many computer languages. This gives us an insight into the developers' hopes for Asterisk. It is designed to be flexible enough to meet any need in the telephony realm.

Second, Asterisk is an open source software package. Hundreds, if not thousands, of developers are working every day on Asterisk, extensions of Asterisk, software for Asterisk, and customized installations of Asterisk. A big portion of the product's flexibility comes from the availability of the source code. This means, we can modify the behavior of Asterisk to meet our needs.

Finally, and most importantly, Asterisk is a framework that allows selection and removal of particular modules, allowing us to create a custom phone system. Asterisk's well-thought-out architecture gives flexibility by allowing us to create custom modules that extend our phone system, or even serve as drop-in replacements for the default modules.

What's new in Asterisk 1.4?

Since the last edition of this book, Asterisk has come out with two major releases—1.4 and 1.6. The new features of Asterisk 1.4 are as follows:

- **Pass through ITU standard T.38 fax calls**: Asterisk now supports the passthrough of fax transmissions to a fax machine.

- **IM support for Jabber and Google Talk**: IM software that supports the Jingle protocol can now be connected to Asterisk.

- **Whisper paging**: This is a new feature of call barging, which allows a user to listen-in on a phone conversation and speak. However, the person listening into the conversation cannot hear the conversation. This feature allows an assistant to talk to someone else in the same office when they're on a call. For example, conveying time-sensitive or important information without the person on the other end hearing what's being said.

- **Improved sound prompts (English, French, and Spanish)**: Digium re-recorded all the sound prompts and included higher quality sound files.

- **Generic jitter buffer**: In the past, the jitter buffer was developed just for the IAX protocol. In this new release, Asterisk now supports other VoIP protocols such as SIP and TDM interfaces.

- **Shared Line Appearance**: This feature mimics the traditional PBX Key Systems, allowing subscribers to share external lines (VoIP, ISDN, PSTN), and also provides status monitoring of the shared line. When a user places an outgoing call using such an appearance, all members belonging to that particular SLA group are notified of this usage. They are also blocked from using this line appearance until the line goes back to idle state or the call is placed on hold.

- **Built-in voicemail system**: In the past, you could either store voicemail as files on the Asterisk server or on an external database. Now voicemail can be retrieved through IMAP on any IMAP-compliant storage system. One benefit of this is unified messaging. This means you can now read a message in your email client and once it is marked read, you will see the **MWI (Message Waiting Indicator)** switched off on your phone.

For a complete list of changes since Asterisk 1.2, visit:
http://svn.digium.com/svn/asterisk/tags/1.4.0/CHANGES.

What's new in Asterisk 1.6?

Most of the changes in Asterisk 1.6 are enhancement changes that improve the reliability and scalability of Asterisk. The new features of Asterisk 1.6 are:

- **New Bridge feature**: In this release, a new Bridge action has been created, which allows a user to connect two existing channels. This functionality will enable the use of advanced features such as in-call announcements and call center monitoring, by a third party.
- **Improved NAT support and support for STUN**: This provides improved connectivity capability with phones located behind a router or firewall.
- **Improved reporting**: A new call event logging capability was developed to give a more complete tracking of events that take place during a call. This will provide more details than traditional **CDR (Call Detail Recording)** and allow more granular tracking and auditing.
- **Support for asynchronous events**: This enables modules in Asterisk to communicate with each other across a cluster. For example, MWI events could be allowed to be distributed among multiple Asterisk servers. This means it is now possible to have SIP endpoints registered to a different server rather than the one holding their mailboxes.

For a complete list of changes since Asterisk 1.4, visit:
`http://svn.digium.com/svn/asterisk/tags/1.6.0/CHANGES`.

Asterisk is a PBX

Asterisk is a Private Branch Exchange (PBX). A **Private Branch Exchange (PBX)** can be thought of as a private phone switchboard connecting to one or more telephones on one side, and usually connecting to one or more telephone lines on the other. This is usually more cost effective than leasing a telephone line for each telephone needed in a business.

Extension-to-Extension calls

First, as a PBX, Asterisk offers extension-to-extension calls. This means users can dial from one phone to another phone. While this seems obvious, elementary phone systems are available (often referred to as Key Systems) that support multiple phones and multiple lines, and allow each phone to use any line.

In operation, the handsets do not have individual extensions that can be dialed, and so there is no way to initiate a call from one handset to another. These systems can usually be identified by having a blinking light for all outgoing lines on every telephone. Unlike Key Systems, Asterisk allows for extension-to-extension calls, allowing directed internal communications.

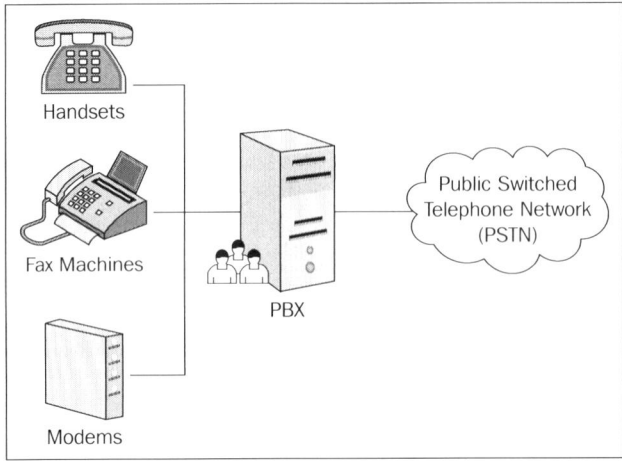

In the previous diagram, each extension (meaning everything to the left of the PBX) can connect to any other extension by dialing it directly. This means if a modem were to send a fax to a local fax machine, it would be done by creating a direct connection between the devices through the PBX.

Line trunking

Secondly, Asterisk offers line trunking. In its simplest form, line trunking simply shares access to multiple telephone lines. These telephone lines are usually used to connect to the global telephone network, known as the **Public Switched Telephone Network (PSTN)**. However, they can also be used as private lines for other phone systems.

These connections can be a single analog trunk, multiple analog trunks, or high-capacity digital connections that allow multiple concurrent calls to be carried on a single connection.

Telco features

Asterisk supports all of the standard features we would expect from any telephone company (or telco). Asterisk supports sending and receiving caller ID and even allows us to route calls based on the caller ID. Using caller ID with the PSTN requires us to subscribe to that feature with our PSTN connection provider.

As expected, Asterisk also supports other features such as call waiting, call return (*69), distinctive ring, transferring calls, call forwarding, and so on. These basic features and more are provided by Asterisk.

Advanced Call Distribution

Asterisk can receive a phone call, look at attributes of the call, and based on that make routing decisions. If enough information is not supplied by our PSTN connection provider, we can ask the caller to input the information using a touch-tone phone.

Once we make a decision on how to route a call, we can send it to a single extension, a group of extensions, a recording, a voicemail box, or even a group of telephone agents who can roam from phone to phone. We can use call queues to serve our customers more effectively while maintaining operational efficiency.

This flexibility gives us the ability to move from having just a phone system, to creating powerful solutions that are accessed through the telephone. **Advanced Call Distribution (ACD)** empowers us to serve our customers in the best way possible.

One major differentiating factor between Asterisk and other PBX systems that support ACD is that Asterisk does not require the purchase of a special license to enable any of these features. For example, the limit on how many calls can be queued at a time is determined only by the hardware we use.

Call Detail Records

Asterisk keeps complete **Call Detail Records (CDR)**. We can store this information in a flat file or preferably a database for efficient look up and storage. Using this information, we can monitor the usage of the Asterisk system, looking for patterns or anomalies that may have an impact on business.

We can compare these records to the bill that the phone company sends out. They allow us to analyze call traffic, say to run a report to find the ten most commonly-dialed phone numbers. We can also determine the exchange that calls us most frequently so that we can target our marketing to the right area.

Moreover, we can look at the time duration of each call. We can count the number of calls a specific agent answers and compare it with the average. There are many uses of this feature.

Using this information, we can also identify abuses of our long-distance calling service. Employees all around the world misuse long-distance call facilities provided by employers. Asterisk gives us the tools to detect possible misuse. The importance of calling records should not be underestimated. This information is invaluable for a variety of business functions. As many countries operate a national do-not-call list, we can quickly determine if we have called anyone on the list to ensure that our verification and checking processes are adequate.

Call recording

Asterisk gives us the ability to record calls that are placed through the PBX. We can use this to provide training material, as examples of calls that went badly or went well. This can also be used to provide call content to satisfy customers or partners, which could potentially be helpful in a legal situation. It's important to consider this feature when setting up your Asterisk service, as you may have substantial hardware and storage issues to address if your PBX is destined to handle and record a substantial number of calls.

Asterisk provides this feature and it is up to us to determine if it is legal, appropriate, and helpful to use in particular circumstances.

Call parking

For users still used to the old Key Systems, call parking is a great feature that allows you to take a call, place it into a parked slot, and then allow another person in the office to pick up that line by accessing the slot. This process mimics the old Key System approach where you pick up a call, place the caller on hold, and then communicate the line number to another person in the office. Instead of a line number, call parking will give an employee a slot number, which if dialed will allow you to pick up that parked call. The slot number will be communicated to the user transferring the caller into call parking, which is accessed by dialing the call parking feature code.

For example, let's say you receive a call in the front office, but you need to check on something in the back. You don't want to transfer the call to the back office because if nobody is there then the caller might end up in voicemail before you reach the phone. Call parking allows you to place the caller into a parked slot. A slot number will be communicated to you. Now you can take your time to go to the back office, pick up a phone, and dial the slot number. Once it is dialed, you will be reconnected with the caller.

Call barging

This is an excellent feature for managers who are training new employees or for those who want to conduct quality assurance. Call barging allows a user to listen to another conversation currently in progress on the Asterisk server. Through Whisper mode, a manager can even communicate to his employee without the remote user hearing the conversation. This allows the manager to coach the employee on a live call without the customer knowing it.

Asterisk is an IVR system

Interactive Voice Response (IVR) revolutionizes just about every business it touches. The power and flexibility of a programmable phone system gives us the ability to respond to our customers in meaningful ways.

We can use Asterisk to provide 24-hour service while reducing the workload for our employees at the same time. Asterisk allows us to play back files, read text, and even retrieve information from a database. This is the type of technology you come across in telephone banking or bill payment systems. When you call your bank, you hear a variety of recordings and issue commands usually using a touch-tone telephone. For example, you may hear greetings and status messages, along with the messages asking you to type in your account number and other personal information or authentication credentials. You will also often hear personalized information such as your last few transactions or your account balance, which will be retrieved from a database. Systems such as this can be and have been implemented using Asterisk.

Asterisk is a call center system

Through the use of queues, call detail records, and its open source nature, Asterisk has become a popular choice among call centers. Queues allow call centers to handle calls in a controlled fashion by placing callers in a holding pattern until an agent is free to take the call. Music on hold can be customized to play messages that further help advertise a company's products or services while the caller is waiting. Other features such as approximate wait time, position in line, and ability to play an IVR with options (such as allowing a caller to leave a voicemail) are some of the enhanced features a call center will need.

Call detail records can also aid call centers as they contain data that can be sorted and put together by queue statistic applications. Some of these open source statistic applications can identify strengths and weaknesses in a call center's routing strategies. For example, the call detail records can record when a caller has hung up and left the queue before an agent has answered the call. This data can be useful as it can identify average wait time and how often callers become impatient and hang up.

Asterisk, being open source has also opened doors for other open source call center applications to be developed for it. For example, today you will find many CRM and predictive dialing applications working with Asterisk.

Asterisk is a voicemail system

Asterisk has a fully-functional voicemail system included. The voicemail system is surprisingly powerful. It supports voicemail contexts so that multiple organizations can be hosted from the same server. It supports different time zones so that users can track when their phone calls come in. It even provides the option to notify the recipient of new messages through email. In fact, we can even attach the message audio.

Asterisk is a Voice over IP (VoIP) system

Asterisk gives us the ability to use the Internet Protocol (IP) for phone calls, in tandem with more traditional telephone technologies.

Choosing to use Asterisk does not mean that we can use only **Voice over Internet Protocol (VoIP)** for calls. In fact, many installations of Asterisk do not use it at all. But each of those systems has the ability to add Voice over IP easily, any time, and with no additional cost.

Most companies have two networks—one for telephones and the other for computers. What if we could merge these two networks? What would the savings be? The biggest savings are realized by reducing the administrative burden for Information Technology staff. We can now have a few experts on computing and networking. As telephony will run on a computer and over our IP network, the same core knowledge will empower our staff to handle the phone system.

We will also realize benefits from decreased equipment purchasing in the long run. Computer equipments get progressively cheaper while proprietary phone systems seem to remain nearly constant in price. Therefore, we may expect the cost for network switches, routers, and other data network equipments to continue to decrease in price.

In most current phone systems, extensions can only be as far away as the maximum cabling length permitted by the telephone system manufacturer. While this seems perfectly reasonable, sometimes we would like it not to be so. When using VoIP we can have multiple users using the same Asterisk service from a variety of locations. We can have users in the local office using PSTN phones or IP phones, we can have remote VoIP users, we can even have entire Asterisk systems operated and run separately but with integrated routing.

One way to slash overhead cost is to reduce the amount of office space required. Many businesses use telecommuting for this purpose. This often creates a problem—which number do we use to reach a telecommuter? Imagine the flexibility if telecommuting employees could simply use the same extension when at home as when in the office or even when using their mobile.

VoIP allows us to have an extension anywhere we have a reasonably fast Internet connection. This means employees can have an extension on the phone system at home if they have a broadband connection. Therefore, they will have access to all of the services provided in the office, such as voicemail, long distance calling, and dialing other employees by extension.

Just as we can bring employees into the PBX from their homes, we can do the same for remote offices. In this way, employees at multiple locations can have consistent features accessed exactly the same way, helping to ease the burden of training employees.

But this is not all that VoIP can give us. We can use an Asterisk server in each office and link them. This means each office can have its own local lines, but office-to-office communications are tunneled over the Internet. The savings to be realized by avoiding call tolls can be significant. But there's more.

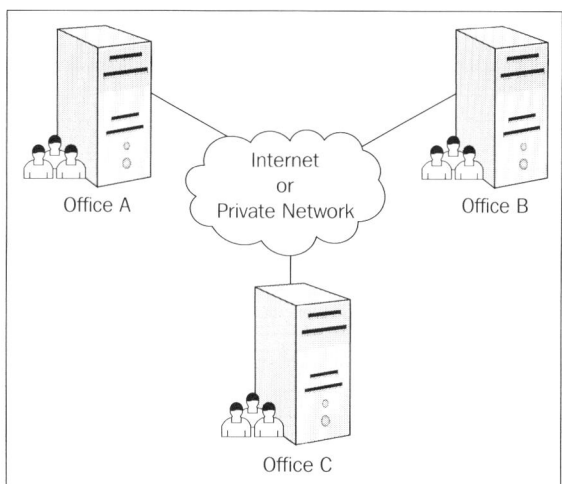

Once we have our offices linked in such a way, we can handle calls seamlessly, irrespective of which office the employees are in. For instance, if a customer calls Office A to ask about their account, and the accounting department is in Office B, we simply transfer the call to the appropriate person in the other office. We don't have to care about where that other office is. As long as they have a reliable Internet connection, they don't even have to be in the same country.

We can route calls based on cost. If it is more cost effective, we can send our calls to another office, where the remote Asterisk server will then connect them with the regular phone network. This is commonly referred to as toll bypass.

Another benefit of linking our phone systems together is that we can route calls based on time. Imagine we have two offices in different time zones. Each office will probably be open at different times. In order to handle our customers effectively, we can transfer calls from a closed office to the one that is open. Again, as we are using an Internet connection to link the offices, there is no additional expense involved in doing so.

By linking our offices together using VoIP, we can increase our customer service while decreasing our expenses—a true win-win situation.

The existence of all these options doesn't necessarily mean we should be using them. With the versatility of Asterisk, we may use or ignore options as it suits our requirements. If we were to use every single line type and feature that Asterisk supports, it could lead to a very complicated and difficult-to-administer system. We should choose the subset that fits our requirements and would function well within our current communications setup.

Asterisk 1-2-3

Setting up Asterisk and working with configuration files without a database is not intended for a beginner. Originally, Asterisk was not considered an off-the-shelf PBX. However, in recent years all of this has changed.

For those who are looking for an off-the-shelf Asterisk PBX system, Digium created the Asterisk Appliance, a feature-rich PBX solution that's easy to install and manage. The Asterisk Appliance allows users to use traditional analog lines as well as a VoIP service provider.

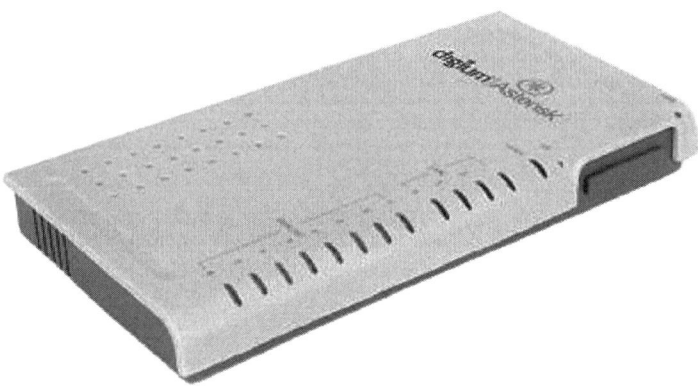

For those who are just beginners, there is a packaged solution called Trixbox CE (www.trixbox.org). Trixbox CE offers a free single CD installation that installs Linux, Asterisk, a database (MySQL), as well as an easy-to-use web-based interface to create and manage your PBX settings. The installation takes approximately 30-60 minutes and once complete, you have a VoIP server ready to go. However, if you want to connect traditional analog lines to your server, you will need to purchase an FXS/FXO card. Please note that for connecting standard **POTS (Plain Old Telephone Service**) lines to your Asterisk PBX, you will need to purchase an FXO expansion card called Fonality.

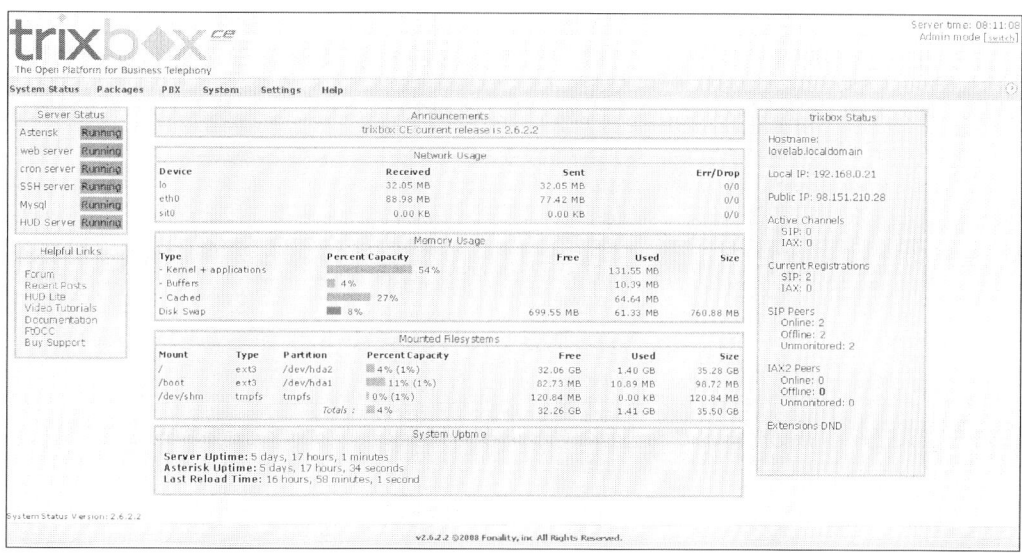

For those of you who are a bit more technically inclined and desire to install each piece of Asterisk individually, you may still want an easy-to-manage interface for your deployment. FreePBX is an easy-to-use GUI (graphical user interface) that controls and manages Asterisk (www.freepbx.org).

Introduction to Asterisk

Another great resource for those interested in FreePBX is the book called *FreePBX 2.5 Powerful Telephony Solutions*. You may also visit:

http://www.packtpub.com/freepbx-2-5-powerful-telephony-solutions

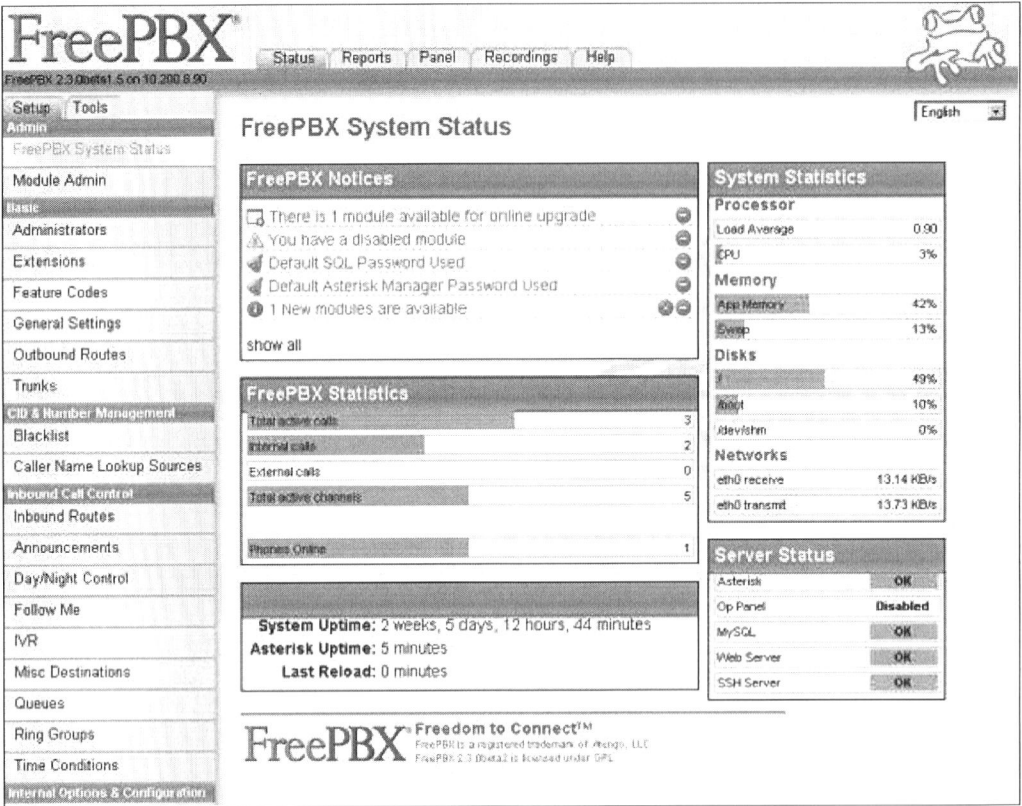

Asterisk scalability

In the past, Asterisk was not a solution for those requiring 100 SIP devices or more. However, in recent years major releases have dramatically increased reliability, scalability, and capacity. Today Asterisk servers can support hundreds of extensions and up to 240 simultaneous calls. For example, Asterisk Business Edition has been tested to handle up to 240 simultaneous calls without any issues. However, it being computerized, the speed, capacity, and reliability is fully dependent on the parts that make up the system. For this reason, ensure you have enough hard drive space, RAM, and CPU power to run your Asterisk server. Those of you who will be using a VoIP service provider for origination (receiving incoming calls) and also termination (outgoing calls) supporting SIP/IAX devices on remote networks, please ensure you have enough bandwidth from your ISP.

Asterisk does not run on Windows

At one point, Asterisk had a demonstration CD that worked with Windows. However, Asterisk offered direct from Digium does not run on the Microsoft platform. Asterisk requires near real-time access to system resources. It also requires hooks into certain resources. Actually, Asterisk is built to use Linux, the open source *NIX operating system.

AsteriskWin32 (http://www.asteriskwin32.com) is an open source project that has managed to get Asterisk 1.2.26.2 compiled for Windows. However, it is highly recommended that you stick with Linux as you will find more support for it in the Asterisk community.

Is Asterisk a good fit for me?

Looking at what Asterisk is and is not, the natural question follows—is Asterisk right for me? This is a vitally important question that should be given serious consideration. Let's take a moment and look at some of the considerations we must explore before we commit to using Asterisk.

Trade-offs

There are a series of trade-offs we must consider with Asterisk. Choosing Asterisk will lock us into certain choices, while others will be available whether we install an Asterisk server or not. We will now examine some of these trade-offs so that we can gauge the impact they have on us.

Flexibility versus ease of use

Asterisk is a very powerful framework into which we can install almost anything. We can configure each piece of Asterisk to the minutest detail. This gives us an amazing amount of flexibility.

This flexibility comes with a price. Each of these details must be researched, understood, and tried. Each change we make affects other parts of the phone system, whether for good or bad. Asterisk is not an easy-to-use platform, especially for a beginner.

There is a learning curve, but it is one that can be surmounted. Many developers have become experts in telephony and many telephony experts have mastered server administration. But each of us must decide what we expect from our phone system. I like to think of it in three major categories, as outlined in the following table:

Description	Solution
I want to plug in the telephone system and never think about it again. I want to call someone when things are not working. I do not plan to add anything to the system once it is set up.	A proprietary phone system is probably your best bet. Many offer a pre-configured system, and when changes are made, a certified consultant will be required.
I don't know much about phone systems, but I want to learn. I need a phone system soon. I'd like to have flexibility and additional features, and may change the configuration of my phone system from time to time.	Either use a packaged version of Asterisk or have a consultant build a customized Asterisk server. Learn to use Asterisk. Build a couple of Asterisk servers just to explore. Add features as necessary.
I want to learn and build my own phone system. I am interested in creating a custom solution for my problems. I am willing to accept the responsibility if something doesn't work, and take the time to figure out why.	Build an Asterisk server from the ground up. Much will be learned in the process, and the result will be an extremely powerful business tool.

Of course, these are not distinct categories. We each fall into a continuum. It is important to realize that Asterisk, as great as it is, is not the right solution for everybody. Like any technology we implement, we must consider its impact on the business. We must also decide whether it will become something useful that enables us to work better, or whether it will require too much maintenance and other work to make it an efficient addition. This depends entirely on our purposes and the other technology we have that requires our attention.

Graphical versus configuration file management

Asterisk currently uses plain text files to configure most options. This is a very simple way to create, back up, and modify configurations for those who are comfortable with command-line tools.

Some PBX systems offer a GUI to update the configurations. Others don't allow the configuration to be changed except by dialing cryptic code on telephone handsets. Still others cannot be configured at all, except by certified technicians who receive the required software and cables from the phone system manufacturer.

A few good open source tools are being created to ease the management of Asterisk. However, to get the full ability to customize Asterisk, editing of text files is still required. To help get used to this method of configuration, this book focuses on the text files without relying on any GUI package.

Calculating total cost of ownership

Asterisk is distributed as free, open source software. The only costs involved with Asterisk are hardware, right? Well, maybe not.

As we have been discussing, Asterisk is very flexible. Determining how to use the flexibility in the best way can quickly become a huge time sink. Compatible handsets are also not free. If we are going to use the G.729 protocol, which compresses VoIP traffic by a factor of eight while maintaining excellent voice quality, we will also have to pay licensing fees.

With commercial phone systems, the costs are typically higher than with Asterisk. However, they are a fixed, known constant. Depending on the way we use Asterisk, costs can vary greatly.

The total cost of owning Asterisk can also include downtime. If we choose to support Asterisk on our own, and have to work to try to get Asterisk back up after a failure, there is an opportunity cost involved in the calls we should have received. This is why we should choose to support our phone system internally only if we have the appropriate resources to back that up.

Total Cost of Ownership (**TCO**) is not an easy calculation to make. It involves assumptions of how many times it will break, how long it will take us to get it up and running, and how much the consultants will charge us if we hire their services.

TCO is useful only when comparing phone systems to each other. The following elements should be included when comparing TCO of multiple phone systems:

- **Procurement cost**: This is the cost to buy the PBX. In the case of Asterisk, it is only the cost of the hardware; other systems will include an element of licensing.
- **Installation cost**: This is the cost to configure and deploy the PBX. Some companies choose to do the deployment in-house. In such instances, there is still a cost, and to enable fair comparisons it should be included.
- **Licensing cost** (one-time): This is the cost of any one-time licensing fees. Some PBX systems will require a license to perform administration, maintenance, connection to a Primary Rate ISDN line (PRI), and so on. In Asterisk, this would include the G.729 licensing cost, if required.

- **Annual support cost**: This is the estimated cost of ongoing maintenance. Of course some assumptions will have to be made. In order to keep the comparison fair, the same assumptions should be carried over between vendors.
- **Annual licensing cost**: Some phone systems will have an annual cost to license the software on the handsets as well as a license to be able to connect those handsets to the PBX.

When we have created the table, we can calculate the TCO for one year, two years, and so on. We can then evaluate our business and decide what costs we're willing to incur for our phone system.

Return on Investment

The cost of owning a phone system is only one piece of the **Return on Investment (ROI)** puzzle. ROI attempts to quantify an expenditure's effect on the bottom line, usually used to justify a large capital outlay.

Just as an example, one phone system that I installed went into an existing business. Its existing phone system had an automated attendant that had the unfortunate habit of hanging up on customers if they pressed the *0* key, or if they didn't press any key for 5 seconds.

What was the ROI for moving to a new phone system? Not having angry customers who got hung up is a hard value to calculate. According to one of the owners of the business, that value was infinite. That made the cost of Asterisk very easy to justify!

ROI is basically the TCO subtracted from the quantification of the benefit (in money) to the business. Therefore, if we calculated that a new phone system would save $5000 and cost $4000, the ROI would be $1000.

Another interesting calculation to make, which is also categorized as ROI, is the time for the cost to be recouped. This calculation is the one that I find helpful in making a business case for Asterisk.

Suppose a phone system costs $5000 to install. Using toll bypass, you can save a net $500 per month. In 10 months, the cost of installing the system will be swallowed up in the savings.

These are simple examples, but ROI can help to justify replacing an existing phone system. By having these numbers prepared before proposing to replace the phone system, we can have a more professional appearance and be more likely to succeed in starting our Asterisk project.

Summary

Asterisk is a powerful and flexible framework, based on open source software. It can be used to create a customized PBX for almost any environment. However, it is not always the best choice for reasons we have just explored. We must consider this carefully in order to be confident that Asterisk is the right choice for our situation. Moreover, we should also ensure that the time and money invested in setting up the Asterisk service is a worthwhile outlay.

2
Making a Plan for Deployment

Now that we have chosen Asterisk to meet our needs, we need to determine our course of action. We will go through some common requirements, discuss the most common choices for solutions, and finally make an informed decision. As we go along, we should make notes to help us on our way.

The Public Switched Telephony Network (PSTN)

Most of the telephones in the world are connected to a vast network, enabling any telephone to reach any other. This network is called the **Public Switched Telephony Network** (**PSTN**). The phones that are on this network are reachable by dialing a number, which may include country codes, area codes, and telephone numbers.

While there are instances in which interconnection with the PSTN is inappropriate, most users of telephones have the expectation that they can reach the world at large. Therefore, we will consider interconnection to the PSTN as a requirement.

Connection methods

There are a number of different methods to connect to the PSTN. Each has advantages and disadvantages, most of which we will touch on. As pricing varies depending on city or country, exact pricing will not be given. Pricing should be researched based upon the location of the Asterisk server.

We will handle each connection method one at a time.

Making a Plan for Deployment

Plain Old Telephone Service (POTS) line

Probably the most common connection to the PSTN is a POTS line. This is an analog line provided by a telephone carrier. Each POTS line can carry only one conversation at a time.

For small installations, POTS lines are usually the most cost effective when connecting directly to our **Local Exchange Carrier** (**LEC**), a term used to refer to any company providing local telephone service. Eight lines is usually the point at which we should seriously look at another technology for our connection.

POTS lines from our LEC require a **Foreign eXchange Office** (**FXO**) interface to be usable in Asterisk. We will focus on Digium's offerings, namely the FXO module on a TDM410. Each TDM410 can use up to four modules. Therefore, if we have one line, we will have three empty module slots on the card.

Integrated Services Digital Network (ISDN)

ISDN is an all-digital network that has been available for over a decade. It is available in two major versions—**Basic Rate Interface** (**BRI**) and **Primary Rate Interface** (**PRI**).

ISDN divides a line into multiple channels. Each channel can contain either payload (Bearing, or B channel) or signaling (Data, or D channel). A BRI has three channels—one D channel and two B channels. Therefore, two phone calls can be in progress at a time on a single BRI. A PRI has 24 channels—one D channel and 23 B channels, resulting in up to 23 simultaneous calls.

ISDN is not limited to voice alone. Each channel can carry 64k of data, if so configured with the LEC. This gives ISDN a lot of flexibility over POTS lines, as the channels can be reconfigured between voice and data on the fly.

With its separate D channel, ISDN is able to do things POTS cannot, such as setting custom caller ID, receiving dialed number information, on-the-fly redirection of calls, and a host of other cool features. Of course, all of these features require cooperation from the LEC, which is not always forthcoming.

BRI does not have high penetration in the United States market. Some accuse LECs of vicious pricing, while others claim consumers are to be blamed for fearing new technology. Either way, the result is the same—if we call our LEC and request a BRI, they will assume it is for data.

On the other hand, PRI is widely used in the US. It is the connection of choice for larger installations. PRIs are actually delivered over T1 connections, a proven and usually very reliable technology.

T1 or E1

Technically speaking, when ordering service from an LEC, we order a DS1, which is delivered over a line referred to as T1. However, this detail is usually overlooked. Therefore, we will refer to it in its vernacular—T1.

A T1 is a line with 24 channels. Each channel can contain a call. Therefore, a T1 can contain an additional call when compared with a PRI. In Europe, E1s are more common. In comparison to T1, they have 32 channels instead of 24. T1s signal the call through Robbed Bit Signaling, also referred to as **CAS (Channel Associated Signaling)** or flat T1. What this means is that a bit is robbed from time to time, as information needs to flow about the connection. While this is usually imperceptible to the human ear, it can be deleterious to data connections.

Using a T1 to deliver both data and voice is common. Some of the 24 channels are designated to be used for data and others are used for voice. There may even be unused channels. LECs are able to offer lower pricing when bundling services in this way, as a few channels may be used for voice, others for an Internet connection, and yet others could be used for a private data connection to another office.

LECs are able to send information about the number that was dialed at the beginning of the call. In this way, one advantage of the PRI has been matched by T1s. If we intend to have about 8 to 12 lines as well as a data connection, a T1 can be a good choice.

An excellent telephony interface card to connect your Asterisk to a T1/E1 connection is the Digium TE122. Today T1 connections can be split to accommodate data and voice. For example, your provider can offer 12 channels of voice as well as a data connection for your computers all on a single T1. The TE122 can support both modes and direct the voice channels to your Asterisk, while separately directing your data connection to the underlying Linux operating system, thus eliminating the need for an external router.

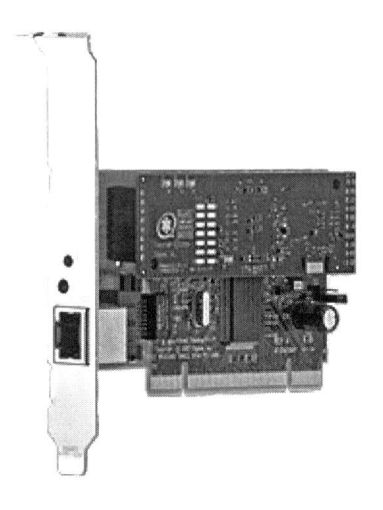

Voice over IP connections

In recent years, a new way to connect to the PSTN has cropped up. Companies are using PRIs, T1, and other technologies to connect to the PSTN, and then reselling those connections to consumers. The users connect to the companies offering these connections through Voice over IP technologies. By doing so, we can skip dealing with LECs completely.

This service is called origination and termination. Through these services, we can receive a real telephone number with the area code, depending on what the provider has access to. Not all providers can offer numbers in every locality. This means that our number could be long distance from our next-door neighbor, yet local to someone in the next state. However, the advantage of this is that the provider will route most of the calls over their VoIP infrastructure and will then use the PSTN when they get to their most local point at the receiving end. This can mean that long distance charges are dramatically reduced. If we call a variety of countries, states, or cities it can be worthwhile to research a provider that offers local PSTN access to the areas we call the most.

The rates per minute are usually very attractive. Often, long distance is at the same rate as local calls. One thing to watch out for is that some providers charge for incoming minutes much like on a cellular telephone, and some providers also charge for local calls.

Today there are VoIP carriers offering unlimited US packages for those running Asterisk. However, one thing to watch out for with unlimited packages is that the carrier usually restricts the number of simultaneous calls you can make or receive. When you inquire about an unlimited package be sure to ask how many channels you are receiving for origination and termination.

Another thing to be aware of is that some providers require you to use their **Analog Terminal Adapter (ATA)**. This means that they will send you a box that you plug into the Internet, which uses Voice over IP. Then, you have a POTS line to connect a phone (or Asterisk) to. However, today many **VSP**s (**VoIP Service Providers**) are offering **BYOD** (**Bring Your Own Device**) in which they provide you with the SIP or IAX settings. Once you have these settings you can connect them to your Asterisk deployment.

Voice over IP makes sense in many installations. But for the quality to be acceptable, a reliable Internet connection with low latency is required. Another thing to watch out for is jitter. Jitter refers to the variation in latency from packet to packet. Most protocols can handle latency a lot better if it is constant throughout the call.

A good candidate for Voice over IP is a site where interruptions in service will not endanger life and will not irreparably harm the company. While VoIP providers strive to achieve very high availability, we also have to rely on the Internet at large and our VoIP provider's ISP, as well as our own ISP.

If our telecommunication needs are such that periodic downtime is tolerable, VoIP will probably be our least expensive option. It requires less hardware in our Asterisk system as well, increasing the savings. In order to use VoIP with Asterisk, all we need is a system capable of Internet access. We don't require any specialized telephony hardware.

Determining our needs

Now that we have examined some of the options, we need to determine what our needs are. Requirements will vary quite a bit from site to site. Something to keep in mind is that, although the previous choices are distinct, they can be mixed in an Asterisk installation. We can have VoIP providers and POTS lines, as well as a PRI if we desire. It's very common to have this type of setup. For example, if we have an office in another country, we can call them using VoIP but all local calls could use POTS. It is important to understand the calls our system will be making and where they will be going, so that we can arrange for the necessary services and ensure that the calls are routed accordingly. If we have an existing telephony system, we can take a look at the calls it's making just now and our current costs so that we can determine what technologies will be of most use to our system's users.

Now is the time to begin documenting what our plan is. If Asterisk is to replace an existing system, then we should start by writing down all the current lines coming into our incumbent PBX. Once that is done, we need to look at our requirements.

First, we need to determine how many lines are needed. Telephone providers can generate a usage report that will tell us the maximum concurrent connections we have experienced in the last month. While they are able to do this, many providers are not very happy to run such a report. However, without that information we have nothing to gauge our needs other than gut feelings.

If we need more channels than we have, someone will get a busy/congested signal. Therefore, we should plan to have the maximum number of channels we have used plus a reasonable cushion. 125 percent of our current maximum is usually a reasonable cushion, this allows for instant 20-25 percent growth so that we can accommodate a sudden increase in calls without the system failing over, causing busy signals. If we do increase calling to this level for a relatively long period, we must consider an increase in lines to prevent congestion. These numbers are a guideline and they can change depending on circumstances. In a call center where the main business purpose is to make and receive calls, 150 percent may be a more satisfactory figure. We also should take into account the time it takes to get new lines set up from our local operator. If a significant event that generates a large number of calls occurs, we should have the capacity to handle this or be able to increase the capacity quickly.

Now that we have a number of lines, we need to determine the technology to use for each line. VoIP is usually the cheapest, especially for long-distance calls. PRI is usually the most reliable, and for incoming calls is often cheaper than VoIP.

While pricing the options, we need to remember that POTS lines usually have a single phone number only, while a PRI can have hundreds of phone numbers. If we are a business that receives only a few calls, but needs the calls to have different phone numbers, then a PRI probably makes the most sense. Also, with a PRI we can trunk more effectively, which may become essential.

Although a PRI can have hundreds of phone numbers, there is a charge for each number each month. Called **DID (Directed Inward Dialing)** numbers, these virtual numbers are usually sold in blocks of 10-20. If we do not order enough to begin with, it is usually not difficult to get new DIDs ordered. Often they can be available the same week, depending on the phone company. We assign these numbers to individual devices or groups of devices ourselves, once we have them allocated.

This means we can decommission or reallocate numbers as necessary. We may have campaign DIDs that are reassigned to different groups depending on the current campaign, personal DIDs for key staff, or our main DID, which would probably be assigned to a group of people responsible for handling these calls.

We should take this opportunity to write down what lines we want, what phone numbers we need, and get quotes if it differs from the currently installed PSTN connections.

Terminal equipment

Now that we have decided on our PSTN interconnection, we need to decide on our internal connections. Our PBX can have modems, fax machines, hardware and software telephones, and other PBXs connected. We will refer to all these different machines as terminal equipment.

Types of terminal devices

There are four major types of terminal equipment—hard phones, soft phones, analog adapters, and PBXs. We will cover each type briefly.

Hard phones

The term hard in hard phones is the short form of hardware. Hardware phones are physical devices that act as a telephone handset. Hard phones are available for POTS (as used in the typical household) or VoIP. Hard phones will typically deliver the highest quality among types of terminal equipment. The most popular hard phones in the market today are:

- Grandstream GXP Series

- Linksys SPA Series

- Aastra 57 Series

- Cisco IP Phones (7940 & 7960)

- Polycom SoundPoint Series

Voice over IP uses various protocols depending on the handset, PBX, and the requirements. The major protocols supported by Asterisk are as follows.

H.323

The first protocol we will be looking at is H.323. Formally known as ITU-T Recommendation H.323, Packet-based multimedia communications systems, this is a suggestion on how to accomplish conferencing over IP, which includes voice, video, and data. This recommendation actually came at about the same time as SIP but has been more widely implemented.

The H.323 standard enjoys full backward compatibility. Currently H.323v5 is out, and v6 is being discussed. Each new release keeps all the pieces of the previous version. This gives a clear upgrade path and some assurance that the equipment won't be quickly antiquated.

H.323 equipment is widely available. From gateways to telephone handsets, all of the needed equipment is relatively easy to find. Most of the telephone handsets are full-featured because the H.323 protocol has a robust feature set.

While the H.323 standard was not designed for wide area networks, a whole set of rules allowing cross-domain addressing have been created. A system for reporting **Quality of Service (QoS)** back to a server has also been developed, allowing such information to be used to route future calls.

Finally, H.323 as a standard supports call intrusion. New endpoints can be added dynamically to any conference (that is, a call) at any time.

Asterisk support for H.323 is not built in. Instead, an additional package, `asterisk-oh323`, must be installed. After installation, H.323 handsets and gateways can be addressed much like any other channel in Asterisk.

SIP

The **Session Initiation Protocol** (**SIP**), is another method of signaling VoIP calls. SIP is a part of the default installation of Asterisk.

Most of the new VoIP equipment supports SIP. SIP has a number of advantages. One such advantage is that the code is smaller. The reason for this is that SIP only supports very basic features. All advanced features are supported through separate Internet standards. Another reason for its small footprint is that, as features are deprecated, the code to implement them is ousted.

Another advantage of SIP's design is its modular nature. As such, extending the protocol is easier to do. It also scales better and was designed with a large network in mind.

SIP seems to be the future of VoIP. There are many features that H.323 has, but are not available on SIP. This includes handset conference control, better Media Gateway definitions, and data sharing. However, SIP is a very good protocol for simple phone calls. Also, as we are using Asterisk, conferences are controlled by Asterisk, not the handsets. Asterisk is a clear Media Gateway, and when used as such, the ambiguity in SIP is not an issue.

IAX

The **Inter-Asterisk eXchange** (**IAX**) protocol is a protocol created by the programmers who brought us Asterisk. Due to the limitations of SIP and H.323, they chose to create a new de facto standard that would allow Asterisk servers to accomplish many things that are simply impossible with the other standards. They also support some features that are extremely difficult to do in SIP and H.323.

First, IAX pierces **Network Address Translation** (**NAT**) easily. Most firewalls and home Internet gateways use NAT, as well as some service providers. SIP and H.323 have worked hard to develop standards to allow them to break through the different types of NAT. However, IAX can work through most NAT devices right out of the box.

IAX is more configurable than the other protocols when dealing with Asterisk. As the source code is available, we can modify it if we so desire, and then submit those changes to be evaluated for inclusion in future versions of Asterisk. As IAX is not currently an Internet standard per se, there is no standard body to work through, allowing more rapid improvement and growth.

IAX supports the trunking of calls. This means that multiple calls can be combined through a single stream. Through the trunking capability, a significant amount of bandwidth can be saved by not having the overhead of multiple streams.

IAX connections between servers support the switch command with which information on how a call is routed can be efficiently shared between Asterisk servers.

IAX supports a large number of codecs. Any codec supported in Asterisk can be used with channels of this type.

As IAX is an Asterisk-created protocol, there are not many handsets and gateways available. However, as time passes, more and more devices are supporting the IAX protocol.

Just as a note, we sometimes see IAX and IAX2 differentiated. IAX2 has been merged into IAX, and IAX has been deprecated. Thus, if a device claims to support IAX2, it should really be supporting IAX.

Soft phones

Much as hard phones are phones implemented in hardware, soft phones are phones implemented in software. Using all the same protocols available to hard phones, soft phones are far less expensive to implement. By using the general-purpose computing resources of a personal computer, the expensive proposition of replacing all telephones in a building can be avoided.

Before going further, we should recognize that most hard phones are in reality soft phones combined with bit of special-purpose hardware. The computing power of a hard phone is not as vast as that of a PC, and unlike a PC, is specially tuned for carrying voice. Thus, we should not dismiss the use of hard phones immediately.

The sound quality experienced on a soft phone will depend greatly on the available resources on the PC, the quality of the software used, and the quality of the data network between the PC and our Asterisk server.

Soft phones will have a hard time being accepted by some users. In addition to the political issue of having people use their computer to talk on the phone, we also have to address disaster planning. If we lose power, keeping a computer up that draws in excess of 400 watts will be far more difficult and costly than keeping power to a hard phone that draws 15 watts, especially for prolonged outages.

The most significant advantage of the soft phone is cost and portability. In most businesses, desks contain a computer and phone at least. If you can remove the phone there is an obvious reduction in hardware costs. There are a variety of soft phone products available and most operating systems come with a basic soft phone package by default. Also the portability aspect of a softphone can be very appealing to companies that have employees who travel a lot. Imagine you're staying in a hotel that offers high speed access. Simply open your laptop, put on your headset, and you're able to make as well as receive calls as if you were in the office. There are also a variety of open source products available. Some companies such as Counterpath (www.counterpath.com) and Zoiper (www.zoiper.com) offer free clients for Windows, Linux, and Mac.

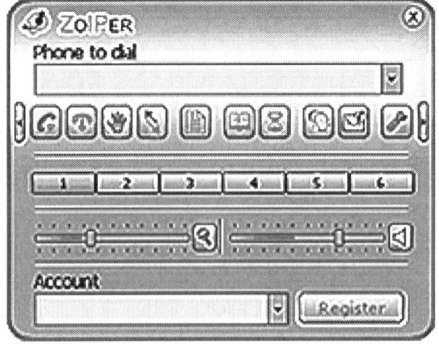

X-Lite 3.0   Zoiper 2.0

If you decide to go with a softphone also be sure to invest in a decent headset with noise cancellation. Headsets that have their own DSP and are USB driven are a good choice. This removes most audio processing resources from the PC so that other work done on the PC does not affect voice quality. The choice of product, soft or hard, is equally important as the PBX. You must be sure that the users will use the device and be sure that it will be reliable and supportable.

Analog adapters

Dedicated communication devices such as modems and fax machines are still very prevalent in business today. Although these devices could be replaced with more modern, reliable, and faster technologies, the new technologies have not yet been embraced.

Therefore, in order to use these existing technologies, analog adapters allow companies to continue using legacy devices such as traditional fax machines. An analog adapter usually consists of an Ethernet jack with which the device will connect to your network and an FXS port (telephone jack) which will connect to your traditional communication device (such as a fax machine).

Another common use of an analog adapter is to connect a cordless phone for those requiring mobility around the office. Granted there are WIFI SIP phones in the market, often users will find that the WIFI signals interfere with other frequency emitting devices. Such interference can cause distortion or the calls to drop.

One issue with analog adapters is their use with traditional fax machines. For years the reliability of faxing over an adapter was poor at best. Today analog adapters are becoming equipped with T.38 capability. This protocol allows regular faxes to be sent over UDP. The use of T.38 dramatically improves the reliability of faxing. However, keep in mind—your VSP as well as your PBX must also support T.38 in order for the fax to be transmitted using this protocol. As of version 1.4, Asterisk now supports T.38 negotiation for SIP users and the related pass through of UDPTL T.38 data. Please note that Asterisk currently cannot terminate T.38 calls or act as a T.38 PSTN gateway without external support.

One extra note about faxing—Asterisk supports receiving and sending faxes through an add-on called SpanDSP. With this Asterisk can receive a fax and turn it into a TIFF file. This TIFF file can then be further processed to become a PostScript or PDF file and be emailed to the proper recipient. Another notable fax detection solution is NVFaxDetect. The installation of this add-on is not covered here, as it is changing rapidly.

These communication devices are usually supported for legacy reasons. We should continually strive to reduce outdated technologies and replace them with up-to-date solutions.

Another PBX

We can connect PBXs together to provide services to users hosted on another PBX. We can use SIP, PRI, T1, H.323, or IAX to connect the PBXs.

If we are connecting multiple Asterisk PBXs, we should use IAX. The IAX protocol has a number of features with this specific use in mind, such as the ability to have multiple conversations trunked into the same UDP stream yielding greater efficiency.

Choosing a device

Now that we have seen the broad offerings of terminal devices, we will see how difficult it can be to choose one to meet our needs. After choosing a type of device, we have to choose a manufacturer and model. This task can be daunting. Let's take a few minutes and discuss how we will make the best decision based on the available information.

Features, features, and more features…

As we review available phone handsets, we will be inundated with all the features that manufacturers can throw at us. These lists are overwhelming, even to the most seasoned experts. It is very difficult to compare two handsets solely on features, as some features have different names.

Determining the usability of a particular phone handset should be a straightforward process. This process has four major steps—requirement elicitation, prioritization, and documentation, followed by handset testing.

Requirement elicitation

This is the brainstorming step. We should go to each user and determine what his or her needs are. We ask the user what features he or she uses on the current phone. We observe the person working for a period of time to get a good sampling of what he or she actually does.

We should then go to the user's manager and see what a person in that position is expected to do. We add these features to our list. While this list will be unique to each user, many will be very similar. We should see patterns of usage emerge between groups of employees.

Requirement prioritization

In this step, we take our requirements list from the previous step and, working with the user and manager, determine which features are used most, which are most important to that user's role in the organization, and which features are simply nice to have. We should also attempt to recognize any deficiencies in the current technology. Changes are often embraced if the change adds value to the user by making a task easier or in some cases removing a task entirely. It's important that we recognize all nuances of the current system in order to provide the user with a replacement that will suit them.

We should then create a quantitative scale for each feature. For example, if we were working with an operator, a Transfer button would be a value of *10*, while a Do Not Disturb button will probably be a value of *1*. If we had a phone handset with both of these abilities, then we add these scores together and it would score an *11*. By putting numbers on the required features, we can come up with a quantitative answer to a very subjective issue.

Requirement documentation

This step is the most important of all of the steps so far, especially for consultants. We take the list of requirements and their weights, and write them in a short document. We then have the user and the manager sign it to indicate their agreement.

This may seem a little formal for picking a telephone handset, but it is an effective method of communicating expectations and plans between you, the implementer, and the users. This can help in preventing surprises or differing recollections of what was promised.

Phone testing

This is the final step. After comparing the available handsets against the document we created in the previous step, we choose the highest scoring handset. We then take a handset of that type to the user and have him or her use it (if we have a test system installed by this point) or at least sign off on it conceptually.

Again, this is an opportunity to ensure our users' expectations are reasonable, that commitments are clearly defined, and that our users are kept informed during the decision-making process. It can also help us get a buy-in from the users as we make the major adjustments that will invariably accompany a new phone system.

Determining true cost

When we look at which handsets to compare our requirements document with, the issue of cost also will have to be looked at. Before we offer a handset that would not be possible under our project budget, we should determine that the handset meets all of the requirements of the business, which includes the element of cost.

The issue of cost is not as simple as looking at the retail price of a handset. Each type of phone will have multiple types of cost. These costs will usually fall into one of the following categories:

- **Handset cost**: This is the easiest cost to determine. It is the actual amount of money that will have to be spent to acquire the telephone.

- **Port cost**: This cost is the element of what the phone connects to on the other end. For instance, on a VoIP phone this could be a portion of the cost of a new network switch that supports Quality of Service (QoS) to enable reliable voice communications.
- **Headset cost**: If a phone will require a headset, then we should consider the cost of that headset as we choose the phone. Different connectors are available depending on the model.
- **Software license cost**: Some phones will require the purchase of G.729 license. Other phones may require a license for the software on the phone (usually referred to as firmware). We should not fail to consider this cost while computing the cost of the phone.
- **Installation cost**: The time required to install different phones differs. This time translates into cost.

By considering each of these factors for each different handset, we get an idea about the true cost of each particular phone. With all of these costs defined, we can see which phones are within our budget and which are simply too expensive.

Compatibility with Asterisk

Not all handsets interoperate equally with Asterisk. Referring to the Asterisk Users Mailing List archive, we can ensure that no serious incompatibilities have been discovered. Also, a Wiki is available at `http://www.voip-info.org`. A vast array of useful information about Asterisk is available there. This site is searchable and is constantly updated.

We do not have to select a single protocol for all VoIP phones. Instead, we can mix and match protocols to our best advantage, thanks to the flexibility and power of Asterisk.

Sound quality analysis

Sound quality is a very subjective thing. Each user will have a personal threshold between acceptable and unacceptable.

Each phone will have a varying sound quality. The variables that can affect the quality of a call are staggering. Network latency can significantly affect sound quality, but so can configurations of the phone. Determining what the cause of low sound quality is can be difficult to do.

Build quality from a manufacturer can also affect quality. When wide variations are allowed from one phone to the next, the result is usually inconsistent from handset to handset. Thus, we have to choose a manufacturer we can trust.

While there is no absolute, the quality of sound on telephone handsets, from highest to lowest, is usually as follows—analog hard phone, VoIP hard phone, analog soft phone, VoIP soft phone. If you are doing a comparison between different handsets, the main things to pay attention to are the amount of background noise (or hiss), distortion, drop outs, popping, and highly digitized voice. If we have users who are extremely sensitive to sound quality, analog will probably be our best bet. For those users who are a little more forgiving, VoIP allows us to use one network for our phones and computers.

When determining what terminal equipment to use, we need to consider the sound quality of each device and match it against the needs and expectations of our users, and weigh that with the cost of that device as compared to the budget.

Usability issues

The world's most advanced VoIP handset is absolutely useless if our users cannot figure out how to use it. As we decide what equipment to provide for our users, we should consider where they are at in the continuum of technological awareness. While VoIP hard phones with context-sensitive buttons are useful for most users, some people find the interface confusing and frustrating.

This is one big issue that we need to address in the handset testing that we do after eliciting the requirements that our user has for a new phone. We have a duty to ensure that our users can use the handsets we choose. We must be careful not to assume that they will figure it out, as doing so often causes hurt feelings and resistance to change. The success of Asterisk will be largely measured by the response of our users.

Recording decisions

It is time to decide what kinds of terminal equipment we will use with Asterisk. First, we should make a list of all users of our phone system. Based on the requirements we get from them and their supervisors, we should decide what type of device to use, whether it is a hard phone or a soft phone. Next, we should determine a protocol to use. Finally, determine a brand and a model of phone to use.

We should take the time to write this down. This list should be provided to the decision makers, and kept up-to-date as changes occur, which they inevitably will. Again, doing so will keep everybody informed and rein in the expectations to keep them reasonable.

How much hardware do I need?

This is probably one of the questions most frequently asked by those who are new to the world of Asterisk. The answer depends largely on what we are going to do with our system.

Conversations that bridge between codecs (called transcoding) take maximum power to handle. Voice over IP conversations seem to take a little more processing power than straight Time-Division Multiple-Access (TDM) calls. Having our server run scripts to find information will take more power than if we define everything statically. How many different conversations we have going at a time will affect how much horsepower we need our server to have as well as the features we use.

Do you see the complexity of answering this question? We have to figure out what we are going to use before we can figure out how big a server we will need. That said, there are some good rules of thumb we can start off with.

First, while we can run an Asterisk server on an old Pentium 90 with 64 MB of RAM, why would we want to? We are creating a robust phone system. We do not have to pay to license the use of the software and we do not have to pay per extension. We can go spend some of the money we saved and buy a decently powerful server. Most would recommend that a small deployment should have a CPU of at least 2.4 GHz and 512 MB RAM; the hard drive space is not as important but typically 120 GB would suffice. The hard drive space greatly depends on how many voicemail messages you want to allow users to store on the server as well as whether you want to record incoming or outgoing conversations. Voicemails and recordings are stored on the server, and without limitation or careful planning you can run out of space. For larger deployments you might want to get a Dual CPU solution as well as an increase in RAM (that is, 2 to 3 GB RAM). As we select the components for our server, we need to remember that we are not building an email server or a web server. We are creating a PBX that people are going to expect to be running all the time. We should select a stable chipset, with an up-to-date BIOS, and match it with other current high-quality components. By using high-quality components, we increase the likelihood of ending up with a high-availability phone system.

On another note, we should select a server with as much redundancy as possible. A RAID-1 controller could save our phone system in the event of a hard drive failure. A pair of RAID-1 controllers that are mirrored could save our phone system in the event of a controller failure or a PCI slot failure. A server with redundant power supplies will help us in the event of power failure or a power supply failure. Of course, our phone system should be on an **Uninterrupted Power Supply** (**UPS**). This is not only for protection from power failures; it will also protect from spikes, and often even lightning.

Depending on the reliability requirements, we might need a redundant server. There are hardware devices that will detect if a PRI is down, and automatically failover. Then again, for most installations, this is overkill.

The most important lesson to keep in mind is that people have grown to depend on phone systems. We should not skimp on hardware, as doing so could cost us dearly in the long run. With the unique pricing structure of Asterisk, all we will have to pay for is any additional hardware to get increased reliability and capacity.

Along with hardware, the question often asked is—which distribution of Linux should I use? If you already have experience with some distribution of Linux, you should be able to make Asterisk work with that distribution. Asterisk is very flexible and has been built with commonly available dependencies, and any distribution of Linux should work. That said, some distributions will require more effort to enable some features such as automatically starting Asterisk when the server boots. As each distribution treats startup scripts differently, most distributions will require a minor amount of tweaking.

Also check the wiki at http://www.voip-info.org for more information on the distribution you intend to use. It has up-to-date notes on compatibility problems, caveats, known issues, and often workarounds for those issues.

Choosing the extension length

While creating our phone system, we will need to create a set of extensions. Although Asterisk has no such requirement, all these extensions should probably have the same length to give comfort to our users. We must determine the length that we will use for all of our extensions.

When creating extensions, it is often advantageous to group certain extensions together. For example, all sales extensions could be in 200s, support in 300s, management in 100s, and so on. Or we could go further and say that all first-tier support will be in 3100s, the second-tier support in 3200s, third-tier support in 3300s, and so on.

We should keep in mind that it is easier to add extensions when there is an available number than it is to renumber all extensions in a building, because we have filled up all of our available dial strings. For instance, suppose we chose 1-digit extensions and have the following phone list:

0 — Operator

1 — Reception desk

2 — Break room

3 — Conference room

4 — John

5 — Sally

6 — Jennifer

7 — Fax machine

8 — Voicemail Access

9 — Outgoing calls

This system will work fine until we add another extension. When we add another extension, we will have to give new extension numbers to all of our users.

Now consider the following phone list:

1000000 — Kitchen

2000000 — Bedroom

3000000 — Office

8000000 — Voicemail

In this house, for someone in the kitchen to call the office (think "Dinner's ready, will you please leave that computer and come eat?!"), the user has to dial seven digits to accomplish what could have been done with one.

Therefore, we need to be smart about how long we make our extensions. Often, if we are replacing a phone system, we should just adopt the numbering already in place to make the transition a little easier for our users. Some phone systems may not have had extension numbers before, such as old analog key systems. All lines were simply visible from all stations. In such instances, we should be sensitive to the new learning that will have to take place and make the length of the extension number as small as possible.

We also need to consider some special instances. First, most people do not want an extension that begins with a 0. Simply put, nobody likes to be a nothing and having a leading 0 for anybody but the operator makes them feel emotionally put down. Also, we should reserve all extensions beginning with a 9 as outgoing telephone calls. Add to that the need to provide services such as call recording, conferencing, and voicemail access. We will give all such services a prefix, such as 8. Thus, we see that we have already lost 30 percent of all of the available extensions.

A good rule of thumb in computing is to take what we believe we will use and triple it, and then round up. Thus, if we believe that at the height of our system, we will have 100 users, we should assume that we will have 300 users. If we believe we will never have more than 10 users, we should assume 30.

With this in mind, here is a table of what we will need:

Expected number of extensions	Our assumed number of extensions	Length to use for extensions
2	7	1
22	70	2
222	700	3
2222	7000	4

Keep in mind when reading this chart that it is much easier to have people dial an extra digit than it is to make them learn all new extension numbers. Thus, if we are a border case, we should go ahead and move on up to the next extension length.

Another idea that we can take advantage of is using an extension that gives a lot of information about the destination. Take for instance, a corporation with seven locations. The first digit in the extension could designate the location. Then the second digit can designate the department, and the remaining digit(s) can designate which member of the group is sought. Thus, knowing the structure and an extension can give an idea of where that person is and what he or she does.

In some environments, such information is not desirable. For instance, in a college campus, some employees work very late at night. If the extension gives their precise location, stalking and threats of physical harm can prove problematic. Therefore, we need to be sensitive to such concerns.

One alternative to these layouts is to use the last few digits of a phone number to refer to each extension. This can work very well if all of such digit strings are unique. However, it can cause problems. Suppose we chose a 4-digit extension and have the phone numbers 555-1234 and 777-1234. Which one is extension 1234? Or suppose we use 7 digit extensions and have (800) 555-1234 and (866) 555-1234. Which one is extension 5551234? Thus, some organizations have moved to a full 10-digit extension length. While this allows 10^{10} extensions, it can cause some users to complain about usability and convenience.

With the flexibility of Asterisk, we can choose many different ways to allocate extensions, all of which will influence our decision on extension length. We must balance our users' expectations along with our desire to leave room to grow. By doing so, we can create extensions that are easy to maintain and user friendly.

Preparing a test environment

Before you deploy a production environment, it is always best to have a test environment available to test settings, try out new software or patches, or simply to practice your Asterisk skills.

For this environment, you can go with an inexpensive solution such as PIII 512 MB PC or better. Another solution is to use a virtual environment in which your PC can virtually host another operating system within itself. An excellent and free virtualization software is VMWARE Server edition (www.vmware.com/products/server/). This software will easily allow you to load ISO files or boot from a CD and install a complete Asterisk system on your local PC.

One thing to keep in mind about test environments is that they are often used for testing functionality. However, quality testing will prove difficult if your test environment is not of the same make as your production environment. For example, virtualization software often requires a great amount of memory and will be sharing system resources with other applications running on your PC. Therefore, calls placed through these deployments may often be degraded in quality.

Summary

Now that we have decided to use Asterisk, we must make a plan. This chapter has looked at the different types of hardware and solutions that an Asterisk system needs, namely:

- What technology we use to connect to the PSTN
- What technology or technologies we use to connect our handsets to Asterisk
- What server hardware we will use
- How we will architect our extensions to be easy to use while also allowing for the growth we can realistically expect
- How to prepare a test environment for Asterisk

As we draw our plan, we must address each of these options before moving on to the next stage, the installation of the Asterisk software, which we will cover in the next chapter.

3
Installing Asterisk

We're making great time! Together, we have selected Asterisk to meet our needs, created a plan to define how our phone system will act, and we are ready to begin installing Asterisk.

Preparing to install Asterisk

In order to install Asterisk, we will need a computer with Linux installed. It's a good idea to ensure your system is up-to-date, for instance, by using the yum tool. Once we have installed our distribution of choice, we need to make sure we have a few additional packages installed. The required extra packages over a base installation are:

- Bison, and associated -devel (1.0.X only) gcc
- Kernel-source
- Libtermcap-devel
- ncurses, and associated -developenssl, and associated -develzlib, and associated -devel

Once we have installed these packages, we are ready to install Asterisk. We should not run an X server or any windowing software on our Asterisk machine, as the resources it consumes are almost guaranteed to delay our voice processing, and therefore negatively impact our sound quality. So, you may save a little time and disk space by choosing not to install any such frontend.

Installing Asterisk

One note here—we should prepare to manage our server. We must keep in mind that we will not be able to rely on graphical tools on the server to manage users, file systems, and other aspects of the day-to-day maintenance that all the systems will need. Unless particularly comfortable with command-line configuration, you should probably consider installing a web-based set of tools such as Webmin, available at www.webmin.com in order to configure Linux. The graphical configuration options for Asterisk that are available are mostly web based. So, we may at some point decide to install these under a web server too, in order to enable graphical configuration.

The commands used to install Webmin are as follows:

`# wget http://voxel.dl.sourceforge.net/sourceforge/webadmin/webmin-1.470-1.noarch.rpm`

`# rpm -U webmin-1.470-1.noarch.rpm`

Once webmin is installed, the default address to access webmin from your browser is `https://<IP of Asterisk server>:10000`.

Obtaining the source files

The very first step we must undertake is to obtain the source files. When obtaining the source code, we have two major choices. We can either download the latest version through Digium's web site from http://www.asterisk.org/downloads or use svn to obtain the latest stable release. The maintainers of Asterisk have been doing a good job of keeping the stable releases available on the FTP servers, so we will use this method.

The commands we issue to download Asterisk's source files are:

`# mkdir -p /usr/src/asterisk`

`# cd /usr/src/asterisk`

`# wget http://downloads.asterisk.org/pub/telephony/asterisk/asterisk-1.6.1-current.tar.gz`

`# wget http://downloads.asterisk.org/pub/telephony/asterisk/asterisk-addons-1.6.1-current.tar.gz`

`# wget http://downloads.asterisk.org/pub/telephony/dahdi-linux-complete/dahdi-linux-complete-current.tar.gz`

`# wget http://downloads.digium.com/pub/libpri/libpri-1.4-current.tar.gz`

The download may take anywhere from about one minute on an extremely fast connection, to a couple of hours on a slow connection. When the download is complete, we will need to unpack the tarballs.

For unpacking the tarballs, you will want to issue a `tar -xvzf [name of file]` (without the brackets):

`tar -zxf dahdi-linux-complete-current.tar.gz`

`tar -zxf asterisk-1.6.1-current.tar.gz`

`tar -zxf libpri-1.4-current.tar.gz`

`tar -zxf asterisk-addons-1.6.1-current.tar.gz`

In the next three sections, we'll compile and install the source distributions we've just downloaded. Note that we should install DAHDI first, then LibPRI, and finally Asterisk.

Installing DAHDI

The DAHDI sources are contained in `/usr/src/dahdi-linux-complete-VERSION` 4. Type the following to install DAHDI:

`cd /usr/src/asterisk/dahdi-linux-complete-VERSION`

`make all`

`make install`

This will take about one to two minutes, depending on the speed of your machine. When it is finished, it should drop us back at the command prompt. If the last message states that there is a failure, we will have to do some detective work in order to determine the cause. The most common issues experienced will be resolved by meeting the dependencies listed earlier in this chapter.

DAHDI contains drivers created for Asterisk that are necessary to use Digium's telephony hardware. It also includes a number of libraries that Asterisk depends on, whether we use Digium's hardware or not.

If we want to have Asterisk start up at boot time, we should issue the command:

`make config`

Installing LibPRI (optional)

If you are using E1 cards, you need to install LibPRI. If you do want to use LibPRI make sure you compile and install it before you compile Asterisk.

Next, we will compile and install the sources contained in `/usr/src/libpri`. We will do this by typing:

```
# cd /usr/src/asterisk/libpri-VERSION
# make clean
# make
# make install
```

This process should take less than a minute. Again, we know it is complete when we are dropped back at the command prompt.

LibPRI provides the libraries required for using **Primary Rate ISDN (PRI)** trunks, as well as a number of other telephony interfaces. Even if we do not have a PRI line at this time, it is a good idea to install it, as it will not create any conflicts.

Parts of the Asterisk code depend on the libraries included in the LibPRI package. Therefore, any time we install LibPRI, we should recompile Asterisk.

Installing Asterisk

Now, it is time to actually install Asterisk, contained in `/usr/src/asterisk`. We will do this by typing:

```
# cd /usr/src/asterisk/asterisk-VERSION
# make clean
```

This installs the runtime and some utilities, as well as libraries of Asterisk PBX. It creates the actual PBX, which may depend on (that is, use) the components we installed earlier.

Through the following commands, Asterisk will now provide a menu in which you can choose which options to install (audio files, voicemail storage, codecs, and so on).

```
# ./configure
# make menuselect
# make
# make install
```

At this point, it is probably wise to install some sample configuration files so that we can acclimatize ourselves to Asterisk's structure. This is done by running:

```
# make samples
```

This creates a sample `DAHDI.conf` in `/etc`, and sample configuration files in `/etc/asterisk`. When we change directories to `/etc/asterisk`, we should see the following files:

- `adsi.conf`: This file contains the configuration for **Analog Display Services Interface** (**ADSI**).
- `adtranvofr.conf`: This file contains the configuration for using Adtran's Voice over Frame Relay.
- `agents.conf`: This file contains the configuration for using agents, as in a call center. This allows us to define agents and assign them IDs and passwords.
- `alarmreceiver.conf`: This file configures the alarm receiver application. We will not be changing the values from their default settings.
- `alsa.conf`: This file contains configuration variables for the console's sound card. We will not be using this.
- `asterisk.adsi`: This file contains Asterisk's default ADSI script. This will be executed from the telephone if we use ADSI hardware.
- `asterisk.conf`: This file sets certain variables for Asterisk's use, most of which we will not need to change. It basically tells Asterisk where to look for certain files and executable programs.
- `cdr_manager.conf`: This file configures CDR for call management.
- `cdr_odbc.conf`: This is the configuration file for using an ODBC database connection to store our CDRs.
- `cdr_pgsql.conf`: This configuration file allows us to use a PostgreSQL database to store our CDR records.
- `cdr_tds.conf`: This is the configuration file for using FreeTDS, allowing connections to Microsoft SQL and Sybase.
- `enum.conf`: This file configures the use of ENUM, which allows us to resolve telephone numbers over DNS. Thereby, allowing us to route calls to an IP instead of going over the Public Switched Telephone Network (PSTN).
- `extconfig.conf`: With this file, we can choose to load our queues through the database engine.
- `extensions.conf`: This file configures the behavior of Asterisk. We will be working with this file extensively.
- `features.conf`: This file contains options for call parking, as well as a few miscellaneous features such as the pickup extension, used for picking calls in each pickup group.

Installing Asterisk

- `festival.conf`: This file sets parameters for Festival, which is an open source program that allows our server to speak text. This is completely optional, and we will not go into configuring this, as Asterisk already includes recordings of the phrases we will need our server to say.
- `iax.conf`: This file configures our Voice over IP (VoIP) conversations using the Inter-Asterisk Exchange protocol (IAX).
- `iaxprov.conf`: This allows for simple provisioning of Digium's S101I, also known as IAXy.
- `indications.conf`: This is where we configure certain behaviors of our phone system, such as ring cadences and tones, enabling us to provide the sounds our users are used to, regardless of what country they are from. We can also mimic their previous phone system.
- `logger.conf`: This file sets up the type of logging we will be using. The defaults work for most people.
- `manager.conf`: This file configures remote access to the Asterisk Call Manager. This will be of utmost importance when we discuss Graphical User Interfaces (GUIs).
- `meetme.conf`: This configuration file sets up simple conference rooms. We can optionally define passwords for the conferences too.
- `mgcp.conf`: This file configures **Media Gateway Control Protocol (MGCP)**. This is a protocol used by some VoIP hardware, mainly from Cisco.
- `modem.conf`: This file sets certain variables to allow us to use selected modems with Asterisk. Please note that not many modems are supported, and as most modems are only half duplex they will not perform very well.
- `modules.conf`: This configuration file selects which Asterisk modules will be started up. We can enable or disable features of our PBX by changing configuration parameters here.
- `musiconhold.conf`: This configuration file creates **Music On Hold (MOH)** instances and defines what music they will play. At this time, it only supports playing MP3s, and only if we install mpg123.
- `osp.conf`: We can configure the Open Settlement Protocol subsystem of Asterisk with this file.
- `oss.conf`: This configuration is much like `alsa.conf`, and we will not be using it.
- `phone.conf`: This file allows us to use some Linux telephony interfaces such as the LineJack by Quicknet. We will be focusing on Digium's hardware offerings instead.

- `privacy.conf`: This file allows us to configure privacy options.
- `queues.conf`: This configuration file allows us to create queues for callers to go through, allowing us to handle burst call volumes in an intelligent way. We can also create an escape to allow callers to dial their way out of the line.
- `res_config_odbc.conf`: This file sets the configuration for storing our settings in an ODBC database.
- `res_odbc.conf`: This is another piece of configuration for storing our settings in an ODBC database.
- `rpt.conf`: This file allows us to use a radio repeater.
- `rtp.conf`: This configuration file sets the ports to use for **Real Time Protocol** (**RTP**). Note that the numbers listed are UDP ports.
- `sip.conf`: This configuration file defines **Session Initiation Protocol** (**SIP**) users and their options. We can also set global options for SIP, such as what address to bind to, what port to use, and what timeouts we are going to impose. SIP is a different protocol for Voice over IP.
- `skinny.conf`: This file configures the skinny VoIP protocol, which is used by many of the Cisco phones.
- `telecordia-1.adsi`: This is another sample ADSI script.
- `voicemail.conf`: This configuration file creates voicemail users and some global options for Comedian Mail, Asterisk's voicemail system.
- `vpb.conf`: This file configures VoiceTronix hardware. We will be focusing on Digium's hardware offerings.
- `chan_dahdi.conf`: This file configures DAHDI telephony interface settings. We will be using this to configure Digium's hardware offerings. Digium's hardware is what allows us to communicate with the PSTN.

As you can see, there are quite a few configuration files. Any particular installation of Asterisk may only use a few of these files, but they are all included so that we have the flexibility to use different features and may expand our services easily.

If we wish for Asterisk to start at boot time, we can configure it to do so by typing:

```
# make config
```

This will create a script to start the Asterisk PBX after the DAHDI startup procedures have been completed. In Red Hat Linux, this script is placed in `/etc/init.d` and set to execute when entering the run level we are currently in (which should be 3).

Getting to know Asterisk

Now that we have installed Asterisk, there are some basic behaviors of Asterisk that we need to explore.

First, the major configuration files that need to be modified are in `/etc/asterisk`, with the exception of `DAHDI.conf`, which is in `/etc`. Each file that ends in `.conf` is a configuration file, which sets parameters for some specific part of Asterisk, as described in the previous section.

The layout of these configuration files is generally simple. Most configuration files will have a variable name, followed by `=>` and its value. For instance, if there were a variable called `cat`, which was to be set equal to Garfield, it would look like this:

```
cat => Garfield
```

In most configuration files, the variables stay set until they are either undefined or set to a new value. You must be careful in what order you set variables, as they may have inherited a different setting than you anticipate. Therefore, I suggest we set all needed values for every single instance, in case we make changes in future that would break a set of lines in our configuration. This will make more sense later.

We should probably take a few minutes and look through all of the `.conf` files in `/etc/asterisk`. Most of the files are commented pretty well. We will be going through the key files step-by-step in the next chapter, but having a general knowledge of what is in there might help us.

If you are planning to use FreePBX or you're using a packaged out of the box software such as Trixbox, it is important not to modify any `.conf` file that is not named `XXX_custom.conf`. If you use either of the above, any changes you make directly in the `.conf` files will be overwritten when you apply your changes in the GUI. However, any file ending in `xxx_custom.conf` will not be overwritten (for example, `extensions_custom.conf`).

Next, how do we start Asterisk? It's really pretty simple. If you set Asterisk to start at boot time, you can reboot. If not, the command to start Asterisk is `asterisk`. When starting Asterisk, there are a number of command-line arguments we can specify. The most commonly used are `-c`, which gives us a console connection, and `-v`, which gives us a verbose output, with more `v`s giving us more information about activities, status, and errors. When I start Asterisk from the command line, I usually use:

```
# asterisk -cvvvvv
```

This gives me a console connection with plenty of debugging information.

If Asterisk starts at boot time, we can reconnect to the Asterisk console by typing `asterisk -r`, and we can specify a level of verbosity. So, when I reconnect, I often use `asterisk -rvvvvv`.

```
[root@pod3 root]# asterisk -rvvvvv
  == Parsing '/etc/asterisk/asterisk.conf': Found
Asterisk CVS-04/26/04-16:26:16, Copyright (C) 1999-2004 Digium.
Written by Mark Spencer <markster@digium.com>
=========================================================================
Connected to Asterisk CVS-04/26/04-16:26:16 currently running on pod3 (pid = 168
0)
Asterisk Ready.
    -- Remote UNIX connection
pod3*CLI> _
```

Although the Asterisk console may not look like much, it is a wonderful tool for checking the status of Asterisk, as well as diagnosing problems. We can type `help` to get a list of commands.

One very useful ability of the console is to issue a reload. This is done by:

`CLI> reload`

This command will re-parse the configuration files and update the changes in most of the modules of Asterisk.

There are some parts of Asterisk that require a restart in order to reflect changes. In order to restart Asterisk, we must first choose when we will be restarting. We have three main choices:

- `now`: This option stops all calls in progress, immediately stops Asterisk, and starts it again.
- `gracefully`: This option does not stop calls in progress, but does not allow any new calls to be started. When all calls that are in progress are completed, the server will restart. Be careful, as a hung channel will basically disable your server.
- `when convenient`: This is my favorite option. This option does not end any calls in progress, and allows new calls to start as usual. When there are no calls in progress, the server will restart. This means that incoming and outgoing calls are not interrupted except for the short period of time in which Asterisk is actually restarting. This would not work, of course, on high-load servers, as there may never be a time when there are no calls in progress. It will also not work if a channel is stuck, meaning the server thinks it has a call in progress, but it does not.

We restart Asterisk using:

`CLI> restart <choice>`

For instance, if we wanted to restart now, we would type:

`CLI> restart now`

Now is a good time to play around with the console interface. Get comfortable with it. Experiment with the *Tab* key to autocomplete commands, and try different verbosity levels to see what information is displayed for each.

I cannot stress enough that now is the time to play with the server. If it is going to be broken by a mistake, it is better that the mistake be made before calls are going through it. If Asterisk stops working, we can go back to the beginning of this chapter and reinstall with minimal time.

Summary

In this chapter, we installed Asterisk and several packages such as DAHDI and LibPRI. The configuration files that we use to set up various aspects of our PBX system have been introduced, as has the Asterisk console. Before our system is ready for use, we need to configure Asterisk for our hardware and other operational requirements, which we'll do in the next chapter.

4
Configuring Asterisk

So far, we have decided to use Asterisk to meet our needs, created a plan, prepared a server, and installed Asterisk and its supporting libraries. Now we have come to the more artistic part of any open source solution—configuration. We get to choose how to use Asterisk's power and flexibility to meet our real-world needs.

While the order in which we proceed makes sense, it is not necessary that we follow it precisely. We can configure the pieces in any order we want. The only issue we may encounter is that if we have not completely configured one part of Asterisk, the PBX may not start, or may start without full capabilities. This is not a real problem, as we are still configuring and will be testing our PBX later.

Thus, if we become unsure about how to proceed in one section, it may be best to move on, and configure the next part. Then, we can go back to where we left off. We often need a little bit of time to digest some information.

What we will now do is step through the different technologies we're going to use, and the configuration files that we need to modify in order to follow the plan that we laid out in Chapter 2, *Making a Plan for Deployment*. Be sure to have the planning worksheets handy so that we are sure what we need to do.

> Before modifying the configuration files, we should make a backup copy of each.

We can create a backup copy of the stock configuration files by changing to the `/etc/asterisk` directory and making a copy of the files. For instance, if we wanted to make a copy of the `chan_dahdi.conf` file and call the copy `default-chan_dahdi.conf`, we would type:

```
$ cp chan_dahdi.conf default-chan_dahdi.conf
```

We could proceed to copy each file in the `/etc/asterisk` directory in this manner. However, if we overlook this backup step, the originals are of course available in Asterisk's source, under the `/usr/src/configs` subdirectory. By saving a copy of the default file, we can make any changes we wish to the configuration files without having to worry about trying to remember what the default file contained. If we have made extensive configuration changes ourselves, we may want to be sure that there is a backup copy of our modifications before we perform any other extensive changes.

DAHDI interfaces

For this section, you need the details of the analog lines and terminals that you set out in Chapter 2. For our DAHDI interfaces, we will be modifying two configuration files: `/etc/dahdi/system.conf` and `/etc/asterisk/chan_dahdi.conf`.

system.conf

As we know, the `system.conf` file is in `/etc/dhadi`. We can modify it in a text editor of our choice.

As we make changes to this file, we will have to force the DAHDI drivers to re-read the configuration files in order to detect the changes. If our system is configured to start the DAHDI hardware at boot time, we can accomplish this by running:

```
$ /etc/init.d/dahdi stop
$ /etc/init.d/dahdi start
```

However, if we decide not to start DAHDI interfaces at boot time, we can implement our changes as we go by running:

```
$ dahdi_cfg
```

In order to get more information, it is often helpful for us to use verbose flags. The more instances of v we use, the more verbose the output will be. So, we may wish to use:

```
$ dahdi_cfg -vvv
```

> Changes to the `system.conf` file will not take effect until we have restarted the drivers.

There are a couple of global options that can be set in the `system.conf` file. First, there's `loadzone`, which defines country-specific preferences, such as what pitch a dial tone should have. Also, there's `defaultzone`, which tells DAHDI which zone to use as default if none is specified. For each of these, the value is the two-letter country code. Supposing we are in the United States, our `system.conf` file could contain:

```
loadzone = us
defaultzone = us
```

Asterisk provides a number of defaults and we can configure additional ones in the `/etc/asterisk/indications.conf` file. Members of the Asterisk community have also contributed sections that we can add for particular countries, which can be found by using a search engine, as the Asterisk users list is archived.

In this file, we have two major classes of devices. First, we have lines. This refers to all of the connecting links we have to the PSTN. Then, we have all of our terminals, which refer to modems, telephone sets, cordless phones, fax machines, war dialers, or any other analog devices we may wish to use.

Lines

Each line can have many different types, and which type we use depends on what services our telephone company is providing. In the United States, two common choices are PRI and POTS lines, also known as analog lines.

In previous chapters, we made a list of all the lines we have coming into our building. We also selected which lines we would be tying to Asterisk. As we are only discussing the configuration of a PBX, we will ignore any lines not tied to Asterisk; however, you should keep the documentation for your future reference.

In the `system.conf` file, we need to identify the signaling we will be using. As these are lines to the PSTN, we will be using FXS signaling. If you remember, we learned that FXO devices are what we need to connect to the PSTN. But here we define the signaling Asterisk is to use, not the type of device. The signaling will be exactly opposite to the device type.

For each Digium channel (that is a port on a TDM400P or X100P), we have to define the following:

```
fxsks=1
```

Note that we are defining three very crucial pieces of information here. First, we specify signaling, which is FXS as this interface is an FXO device. Secondly, we designate the protocol. This depends on the phone line that we use. Usually, we can use Koolstart (ks). Other options are Loopstart (ls) and Groundstart (gs). Finally, we set the channel number to 1. This depends on the order in which our system detects the modules. Thankfully, if we configure the values incorrectly and run `ztcfg`, we will see an error message like the following:

```
$ dahdi_cfg
dahdi_chanconfig failed on channel 1: Invalid argument (22)
Did you forget that FXS interfaces are configured with FXO signalling
and that FXO interfaces use FXS signalling?
```

As we can derive from the error message, this is not an uncommon problem. It is no big deal, we just know that our system believes that channel 1 has the opposite type of signaling to what we defined. We can go back into our `system.conf` and switch the signaling. This can help us determine the order in which our system is loading the DAHDI interfaces if we are unsure.

If we have a PRI or a T1 coming in, either from the telephone company or a channel bank, then our configuration will be a bit different. We define each T1 as being a span. For each of these spans, we need to specify the following details:

- **Span number**: This is an arbitrary number that we assign to each T1. The T1 connected to the first port should be 1, the second T1 connected should be 2, and so on.
- **Timing**: This is an integer that represents the order in which we rely on it for synchronizing our timing. 0 means that we will not use this span for synchronization, 1 indicates it is our primary source, 2 indicates that it is our first backup timing source, 3 indicates it is our second backup timing source, and so on.
- **Line Build Out (LBO)**: This integer represents the length of the cable. Valid options are:
    - 0 db or 0-133 feet (if you are unsure, try this)
    - 133-266 feet
    - 266-399 feet
    - 399-533 feet
    - 533-655 feet
    - -7.5 db
    - -15 db
    - -22.5 db

- **Framing**: This information is specific to our connection. While we should be able to get this information from our telephony provider, many users have reported difficulty in finding anybody at the telephone company willing and able to provide this information. Therefore, it is convenient that there are only a few options.

    For a T1, the valid options are:
    - `d4`, also known as `superframe` or `sf`
    - `esf`

    For an E1, the possibilities are:
    - `cas`
    - `css`

- **Coding**: This information is also specific to our particular line, and again should be available from our connection provider.

    For a T1, it could be:
    - `ami`
    - `b8zs`

    For an E1, we could have:
    - `ami`
    - `hdb3`
    - `crc4`: added to `ami` or `hdb3` to enable CRC-4 checking

So, supposing we have a PRI in the US, using ESF framing and B8ZS coding, we would have:

```
span=1,1,0,esf,b8zs
```

Note that generally, D4 and AMI go together, as well as ESF and B8ZS go together. Deviations from these pairings are extremely rare and often based on misconfiguration. Also, if you know that your circuit is a PRI, then most likely the framing and coding are ESF and B8ZS. This provides a few rules of thumb when you have a circuit you cannot get any information on.

Now that we have defined the span, we must configure channels for use. In order to do this, we use statements of the form `<device>=<channel>`. Here, `<channel>` can be a single number representing a specified channel, a comma-separated list of channels, or a range, using a hyphen between the first and last channels. For a T1, some common options for `<device>` are:

- `e&m`: This will work irrespective of implementation, such as Immediate, Wink, or Feature Group D.
- `fxsls`: This is used with many channel banks, referred to as "loopstart".
- `fxsgs`: This is used with many channel banks as well, referred to as "groundstart".
- `unused`: This tells DAHDI to ignore the channel, for instance when only a fractional T1 is delivered by the telephone company.
- `bchan`: Also known as "indclear". This tells DAHDI not to perform any conversion.
- `dchan`: Also known as "fcshdlc". This tells DAHDI to perform HDLC encoding and decoding on the bundle, and send it through this device.

A full list of the possible devices is contained in the sample configuration file, or in `/usr/src/dahdi/system.conf.sample`. Continuing with our example of a PRI in the US, we would have:

```
bchan=1-23
dchan=24
```

If we had two identical PRI lines coming in, the entire `system.conf` file would look like the following:

```
# First incoming PRI
span=1,1,0,esf,b8zs
bchan=1-23
dchan=24

# Second incoming PRI
span=2,2,0,esf,b8zs
bchan=25-47
dchan=48

loadzone=us
defaultzone=us
```

The configuration for a channel bank will be very similar. Supposing we had a single channel bank using ESF framing and B8ZS coding, and Groundstart lines on channels 49, 50, 51, and 52, then we would have:

```
# Channel bank to PSTN
span=3,0,0,esf,b8zs
fxsks=49-52
```

Terminals

Just as our configuration of the lines depends on the capabilities of our telephone provider, as we configure our terminals, we must keep our equipment in mind. Our settings in /etc/dahdi/system.conf for our terminal devices will be directly related to the type of equipment we are using.

As our analog terminal devices will be using an FXS device to connect, Asterisk must use FXO signaling to communicate. Therefore, if we have a TDM400P with FXS modules, we would use fxoks signaling. Assuming this is our second port to load during ztcfg, the entry in system.conf will be:

```
fxoks=2
```

We can also connect our Asterisk server to a channel bank or another phone system. This is done just as it was for defining lines. We must first designate a span (in the same format as before) and then configure the individual channels.

For instance, suppose we have a channel bank that has 4 FXO ports and 20 FXS ports. Imagine it is using ESF framing and B8ZS coding. Thus, we could have:

```
# Channel bank to PSTN and Terminal Devices
span=3,0,0,esf,b8zs
fxsks=49-52
fxoks=53-72
```

No problem, right? Remember that as we make mistakes, when we run dahdi_cfg we will get some useful error messages. The more verbose the output, the more likely we are to get a hint as to what is causing any of the problems we may experience.

chan_dahdi.conf

Now that we have configured /etc/dahdi/system.conf, we have our telephony devices starting. If our FXO or FXS device is not starting yet, we need to go back to the previous section, as this is one of the few configuration files that depend on the success of another. Until we have our interfaces loading, we will not have much success in /etc/asterisk/chan_dahdi.conf.

Configuring Asterisk

This configuration file is read by Asterisk. Therefore, to read changes made to this file, we can issue a `reload` in the Asterisk console. DAHDI will *not* have to be restarted to apply any changes we make in `chan_dahdi.conf`.

At the top of our file, we will see a `[channels]` section. In fact, that is the only section we will have. At the beginning, we will set certain characteristics that we want to be consistent between all of our lines and terminals. We must exercise great care when working with this file. When we set any variable, it stays in force during all of the later channel declarations until we set it otherwise. For those of us who program, it is much like a `switch` statement without a `break` statement at the end of each `case` — sometimes useful, always dangerous!

Each of these settings that we discuss can be reset at any point in the file, whether for incoming or outgoing interfaces. As we reset it while defining one interface, all of the following lines will have the same setting. Therefore, it behooves us to group lines of similar configuration together so that we won't need to reset variables as often.

Not every setting will be needed for every channel. In fact, most installations probably don't even use half of the settings available. However, we should go over all of the possibilities, as every installation is a little bit different, and one of these settings might be just what we need.

As there are so many options that must be set, we will go through them in list format as follows:

- `Language`: This is the default language to use. The default is `en` for English.
- `context`: This is how we tell Asterisk which context to put new calls in. The default value is `default`.
- `switchtype`: This is used only for PRI. The valid options are:
    - `4ess`: AT&T's 4ESS protocol
    - `5ess`: Lucent's 5ESS protocol
    - `dms100`: Nortel DMS100
    - `euroisdn`: EuroISDN
    - `national`: National ISDN 2 (default)
    - `ni1`: Old National ISDN 1
- `pridialplan`: This setting is only occasionally used for PRI connections. The options are:
    - `unknown`
    - `private`
    - `local`
    - `national`
    - `international`

- `overlapdial`: This setting decides whether or not to send overlapping digits while dialing. Valid options are "yes" and "no". However, please note that most PRI implementations will use enbloc dialing. Hence, its best to leave this as "no" unless instructed otherwise.
- `signalling`: This setting chooses the signaling method. Valid options are:
    - `em`: E&M signaling
    - `em_w`: E&M Wink
    - `featd`: Feature Group D, Adtran style
    - `featdmf`: Feature Group D, US
    - `featb`: Feature Group B, US
    - `fxY-ZZ`, where `Y` can be `o` or `s` (this will be the same as in `system.conf`), and `ZZ` can be `ks` for Koolstart, `gs` for Groundstart, or `ls` for Loopstart
    - `pri_cpe`: PRI signaling, Customer Premises Equipment (CPE) side
    - `pri_net`: PRI signaling, Network side
    - `sf`: SF Signaling
    - `sf_w`: SF Wink
    - `sf_featd`: SF Feature Group D, Adtran style
    - `sf_featdmf`: SF Feature Group D, US
    - `sf_featb`: SF Feature Group B
- `prewink`: Prewink time, defaults to 50 ms.
- `preflash`: Preflash time, defaults to 50 ms.
- `wink`: Wink time, defaults to 150 ms.
- `flash`: Flash time, defaults to 750 ms.
- `start`: Start time, defaults to 1500 ms.
- `rxwink`: Receiver wink time, defaults to 300 ms.
- `rxflash`: Receiver flash time, defaults to 1250 ms.
- `debounce`: Debounce, defaults to 600 ms.
- `usedistinctiveringdetection`: Set this to "yes" for our FXO interface if our phone company sends us a distinctive ring.
- `usecallerid`: Set this to "yes" if we wish to use caller ID. Note that this may cause a delay between rings and the pickup of calls by Asterisk, as Asterisk will have to wait for caller ID information to be available. If our installation cannot handle this, then we will have to set it to "no".

- `hidecallerid`: Set this to "yes", if we wish to mask our caller ID on outgoing phone calls.
- `callwaiting`: Set this to "yes" to enable call waiting on FXO devices.
- `restrictcid`: This sends caller ID as **Automatic Number Identification (ANI)** only, and is not available for the user.
- `usecallingpres`: This toggles whether we want to use the caller ID presentation for outgoing calls that the calling switch is sending.
- `callwaitingcallerid`: This sets if we support caller ID for call waiting.
- `threewaycalling`: This sets if we support three-way calling.
- `transfer`: Here we can decide if we're supporting flash-hook transfers. The use of this feature requires three-way calling.
- `cancallforward`: Set this to "yes" if we want to be able to forward calls.
- `callreturn`: Whether or not to support *69 for call return.
- `mailbox`: Here we set the voicemail number of the mailbox. If this is set to a valid voicemail number, and that account has a new message, then our user will hear a stutter dial tone when he or she picks up the phone. If there is more than just one context for voicemail, we can specify it as `user@context`.
- `echocancel`: This variable can be set to "yes", "no", or a number defining how many taps to cancellation (needs to be a power of 2).
- `echocancelwhenbridged`: This sets whether we want to cancel echo when the circuit is purely **Time Division Multiplexing (TDM)**. Usually, echo canceling is unnecessary; however, the sample configuration has it set to "yes".
- `echotraining`: This variable can be "yes", "no", or "800". It has been reported on the user's list that a setting of 800 fixes most of the echo experienced on TDM400P and X100P interfaces.
- `relaxdtmf`: This setting can help if Asterisk keeps detecting voice as DTMF (digits).
- Gain settings:
    - `txgain`: In dB. This sets the TX gain, default of 0.0.
    - `rxgain`: In dB. This sets the RX gain, default of 0.0.
- `group`: We can assign groups to our lines. This allows rollover when making outgoing calls.
- `callgroup`: This is a ring group.
- `pickupgroup`: This is a pickup group. If another phone in your pickup group rings, you can pick it up by dialing *8#.

- `immediate`: This sets if we want calls picked up immediately, or if we want to provide a dial tone. The default is "no", which means that we will provide a dial tone, read digits, and complete the call.
- `callerid`: We can set the caller ID number to "asreceived" or override it with a specific number. "asreceived" only makes sense on trunk lines.
- `amaflags`: This affects the recording of CDRs. It can be set to "default", "omit", "billing", or "documentation".
- `accountcode`: We can tie channels to account codes for billing purposes.
- `adsi`: We can set this to "yes", if we have ADSI-compatible handsets.
- `busydetect`: This can be set to "yes" to try to find whether lines are hung up or busy.
- `busycount`: This variable sets how many busy tones to hear before hanging up a channel. The default value is 4, but increasing this may prevent some seemingly random hang ups.
- `callprogress`: This variable sets whether or not to use the call progress detection algorithms.
- `progzone`: Used in conjunction with `callprogress`. Set this to "us" for the United States.
- `musiconhold`: Sets which class of music on hold to use.
- PRI idle extension: This group of settings can be used to more effectively use channels on a PRI line. It is often used to multilink through PPP.
- `idledial`: Sets the extension to dial from the idle line.
- `idleext`: Sets the extension to dump the idle line to.
- `minunused`: Sets minimum number of channels to leave unused.
- `minidle`: Sets minimum number of channels to leave in the idle extension.
- `jitterbuffers`: Default is 4. This is designed to smooth out jitter.
- `cadence`: Defining custom ring cadences. Defining any here will cause the default cadences to be turned off.
- `channel`: This can be a channel or range of channels. These channels must match those defined in `/etc/dahdi/system.conf`.

While there are many options available, we will only need a few of them. We can set most of them once without the need to set them again. Usually, the defaults work fine, but at least we know that if we're unhappy with the way our phone system acts, we can easily configure it to perform the way we wish it to.

For `/etc/asterisk/chan_dahdi.conf`, we have the two major divisions of devices, as we had in `system.conf`, namely lines and terminals. The definitions are the same as before.

Lines

Once we have set the previous options, all we have to do is define the channel number. It is best to set the signaling method within sight of the channel definition so that problems are easier to debug. Supposing we have an FXO device on channel 1, which we want to call group 1, we would have the following lines in our `/etc/asterisk/chan_dahdi.conf` file:

```
signalling=fxs_ks
context=default
group=1
channel=1
```

Grouping these four lines close together in the file will make it easier to troubleshoot any problems in the future.

Suppose we have two PRIs, as we did in the example in the *Lines* section of *system.conf*. Our channel definitions could look like the following:

```
; incoming PRI
callerid=asreceived
context=default
switchtype=dms100
signalling=pri_cpe
group=2
channel=>1-23
channel=>25-47
```

By putting both of the incoming PRIs together, we saved ourselves some typing. As long as we're happy with both trunks being in the same group, there is no reason why we should have to redefine the variables.

One note on security is that we need to be careful what context we put our incoming calls into. If we place them in a context that can dial long distance, then people can relay telephone calls through our server.

Terminals

When defining the channels that our terminal devices connect to, we need to remember to take into consideration more specific details about the uses of the channel. For instance, analog handsets will be unable to send caller ID information, so setting `callerid=asreceived` does not make much sense.

Furthermore, here is where we have to remember if we are using ADSI devices. While it may be tempting for us to turn on `adsi=yes` for all lines, it tends to make users angry when they hear some of the high-pitched beeps that sometimes are emitted during conversations. Therefore, we should only enable ADSI on phones that are ADSI compatible.

An example configuration for an FXS device on channel 2 might be:

```
signalling=fxo_ks
context=longdistance
callerid="My Name Here"<(850) 555-5555>
adsi=no
callgroup=1
pickupgroup=1
channel=2
```

As you can see, this phone does not support ADSI, has the correct caller ID set, and has been placed in the call group 1, so that the user can pick up any calls for terminals in the same group. Also, we need to take notice that the line `channel=2` appears last.

We can repeat this process for all of the channels we need to define. Lines through channel banks will be configured the same. In keeping with our previous example, suppose we have a channel bank with 4 FXO devices, and 20 FXS devices. The section would look like the following:

```
; channel bank configuration
; fxo devices
signalling=fxs_ks
context=default
group=1
callerid=asreceived
channel=49-52
;fxs devices
signalling=fxo_ks
context=longdistance
group=
callerid="John"<1234>
channel=53
callerid="Jacob"<2345>
channel=54
callerid="Jingleheimer"<3456>
channel=55
callerid="Smith"<4567>
channel=56
callerid="My Company"<(555) 555-5555>
channel=57-72
```

Configuring Asterisk

Notice that I have not given a value for `group` in the `longdistance` context. This moves these channels to be outside a group. In this instance, I don't feel using a group number would be appropriate, as each phone may be for different functions, and so on.

As you can see, the only different piece of information for each of the lines in the channel bank is the caller ID. This makes it very convenient to define all of the channels. We now have plenty of information to configure all of our DAHDI terminals.

SIP interfaces

Session Initiation Protocol, or SIP, is a standardized Voice over IP (VoIP) protocol. This protocol relies heavily on the RTP, which uses UDP ports in the TCP/IP stack. It presents addresses in much the same format as email, as `user@domain`.

We configure this protocol by editing `/etc/asterisk/sip.conf`. This file has a number of settings in a `[general]` section, followed our definitions of users.

There is a whole host of options that we can set. These options include:

- `context`: Sets the default context for calls coming into the server. These calls can be from our users, or if we are connected to the Internet, they can be from anywhere. Just to be on the safe side, we should not set this to be a context that can call long distance. The default is "default".

- `realm`: Sets the realm of the server. As we discussed earlier, the calls are addressed like email, in that the format is `user@domain`. This variable is how we set the domain part. This could be your host name or a domain name. If we enter nothing, it will work, and will set our realm to "asterisk", but we really should set this to our domain name.

- `port`: Sets the UDP port to listen on for connections. The default is 5060, and we shouldn't change this unless we have a *really* good reason.

- `bindaddr`: Specifies the IP address that we want the SIP service to bind to. Keep in mind that if a machine has multiple IP addresses, we can specify that it binds to all by using 0.0.0.0 as the `bindaddr`. Also keep in mind that if we ever change the IP of the server, we will have to adjust this variable, unless we use the 0.0.0.0 address. It may be a good idea to only bind where you expect SIP traffic, for added security.

- `srvlookup`: Determines if DNS SRV lookups are enabled. SRV records in DNS are a way to allow other Internet users to point to our SIP server without having to know its host name. This also allows us to be able to change SIP servers without updating everybody's address. We can think of this in the same vein as a DNS MX record. We should set this to "yes".

- `pedantic`: If we have Pingtel phones, we should set this to "yes". This enables pedantic checking for multiline formatted headers.

- `tos`: This is the QoS setting. We can choose to specify a numerical value, or use the keywords "lowdelay", "throughput", "reliability", "mincost", or "none". We should probably set this to "lowdelay", as users are typically pretty annoyed by pops, cracks, and other inconsistencies in voice conversations.

- `maxexpiry`: This is the maximum length in seconds that we allow incoming registrations to stay valid. We should set this to something reasonable. The default is "3600", meaning registrations will time out after one hour.

- `defaultexpiry`: This is the default length in seconds that we set for incoming registrations. The default is "120", or 2 minutes.

- `notifymimetype`: We can override the MIME type in SIP `NOTIFY` messages. We should not modify this unless we absolutely have to.

- `videosupport`: We can set this to "yes" to allow video support in SIP. Asterisk is currently known to support H.261 and H.263 video; however, since Asterisk 1.4, H.263p and H.264 are also now supported.

- `musicclass`: Here we can set the default music-on-hold class for all SIP calls. The default is "default".

- `amaflags`: We have the same choices here as for `amaflags` in `chan_dahdi.conf`—"default", "omit", "billing", or "documentation".

- `accountcode`: We can set the account code to use for calls from this SIP user. This can help with billing.

- `language`: This is the default language for SIP users. We can set this to that of the country where the server is, and then override this setting for any users with a different language. The default is "en".

- `relaxdtmf`: Just as in the `chan_dahdi.conf` file, we can relax the handling of DTMF in SIP. This can be useful if Asterisk is detecting digits pressed in the middle of voice conversations.

- `rtptimeout`: Here we configure how long a period of inactivity will time out calls. We set this to the number of seconds the RTP stream has to have no activity before Asterisk will terminate the call. The default setting is 60.

- `rtpholdtimeout`: Here we can configure the hold timeout in seconds. We set this to a value greater than the `rtptimeout`. The default setting is 300.

- `externip`: Here we set the external IP address of the Asterisk server. This is very useful for piercing through NAT and some firewalls. This address will be placed on outbound SIP messages.

- `localnet`: Here we define all of the internal IP addresses. This tells Asterisk which SIP messages to use the external IP on, and which ones need the internal IP address. You will list the address range in the format `<network>/<subnet>`. For example, 192.168.1.0/255.255.255.0.

- `register`: Here we can tell Asterisk to register with a SIP provider or service. An example of this would be: `register => john:johnspassword@sipprovider.com`. Optionally, we can specify the port of the remote server to use (if it is different from the default 5060) and the extension to drop all calls into. If we were to specify these, and we wanted to use port 5061 and place all incoming calls into extension 9999, it would look like: `register => john:johnspassword@sipprovider.com:5061/9999`.

- `codecs`: Now we must define what encoders/decoders (codecs) we will allow. This is done through a series of `allow` and `disallow` statements. First, we should disallow all codecs by using the statement `disallow=all`. Then we can enable codecs one at a time by typing `allow=ulaw`, `allow=ilbc`, and so on, in order to allow all of the codecs we wish to use. The order in which we enable the codecs will be the order in which Asterisk tries to negotiate them.

Now that we have configured the global options, we must define our users. As we do so, we can use most of the previous settings, as well as a few new options. Note that in the general section, the order of allowed codecs mattered; however, as we define our users, order will not guarantee the order of negotiation. The following is a list of the options available to us as we define users:

- `type`: There are three major types of users:
    - `user`: This connection is only permitted to send calls to us.
    - `peer`: We are permitted to send calls to this connection.
    - `friend`: This connection is both a user and a peer.
- `username`: Sets the username for authentication.
- `secret`: Sets the password used for authentication.
- `md5secret`: MD5 hash of <user>:asterisk:<secret> for more secure authentication.
- `fromuser`: Overrides caller ID, and is required by **Free World Dialup** (**FWD**).
- `callerid`: Sets the caller ID. An example would be: `callerid=My Name <1234>`. It's usually a good idea to set `fromuser` and `callerid` to the same string.
- `host`: Sets the host address of the user. This can either be a static address or the keyword "dynamic".

- `defaultip`: Used with `host=dynamic`. This sets the default IP for when the extension has not registered.
- `nat`: Sets whether or not this SIP device is behind a NAT firewall.
- `mailbox`: Sets the mailboxes to check for messages for this user. This can be just the mailbox ID, or the mailbox ID followed by `@contextname`.
- `qualify`: How many milliseconds the device can be unreachable before considering it as down.
- `canreinvite`: We should set this to "no" if one of the devices is behind a NAT. When two devices start having a conversation, they try to reinvite each other, thereby skipping the Asterisk server. This is good to keep the load down, as well as to lower the number of hops required in the network, but if Asterisk is how we get through the NAT, the reinvite may not work.
- `call-limit` (in older asterisk versions this is called `incominglimit`): We can set the maximum number of calls a device can initiate at a time. Setting this to "1" will disable the ability to do three-way calls or transfers on some SIP phones.
- `dtmfmode`: This sets the DTMF mode. The choice we make depends on the hardware we're using. The valid options are "rfc2833", "info", and "inband".
- `callgroup` and `pickupgroup`: These can be set as they were in `chan_dahdi.conf`.
- Security: We can use the keywords "permit" and "deny" to provide some level of security. Order matters, as the last matching rule will be the one followed.
- `deny`: Lists IP addresses to deny. If we issue a `deny=0.0.0.0/0.0.0.0`, then all attempts are denied. If we issue a `deny=192.168.1.0/255.255.255.0`, then all connection attempts from the class C space of 192.168.1.0 are denied.
- `permit`: Lists IP addresses to permit. By default, all IP addresses are permitted. If we used deny 0.0.0.0, then we will now have to list all IPs and blocks of IPs that we should allow.

Now that we have gone through the options, let's look at some examples. Suppose we have already set all of the general options. We have three users—one for all incoming SIP calls, one for calling out of FWD, and one for a SIP handset.

```
[sip_incoming]
type=user

[FWD-out]
type=peer
secret=mypassword
```

```
username=myusername
fromuser=myusername
host=host.provider.com

[1000] ;this is the extension of the handset
type=friend
context=longdistance
username=1000
callerid=My Name <1000>
host=dynamic
defaultip=192.168.1.100
secret=2manysecrets
nat=no
canreinvite=yes
dtmfmode=info
outgoinglimit=1
incominglimit=2
mailbox=1000
disallow=all
allow=ulaw
amaflags=default
accountcode=company123
```

As you can see, the definitions can be as short as one line (assuming we defined our default context in the general section) or as long as we need them to be. Now, we need to edit our SIP configuration file for our needs. Take a look at the terminal devices that we listed in Chapter 2, *Making a Plan for Deployment* where we selected SIP as our protocol. We now have enough knowledge to be able to create a user for each one of them.

IAX interfaces

Asterisk provides another VoIP protocol, much like SIP, called Inter-Asterisk eXchange, or IAX. This protocol is easier to work with for many reasons, as we discussed in Chapter 2. For this section, we will need all of the terminal device details for which we selected IAX as the protocol.

Just as the name suggests, IAX is well-suited to connecting multiple Asterisk servers together. At this point in our work with Asterisk, this is not what we need, as we are limiting ourselves to one server. When we link multiple servers together, this feature will become useful.

Following the pattern established, the IAX protocol is configured in `/etc/asterisk/iax.conf`. Just like the SIP file, it has a number of general settings, followed by settings for each individual user.

First, we will discuss options for the `[general]` section. They include:

- `port`: Set the port to listen on. The default is 4569, which we should not change without a very good reason.
- `bindaddr`: Set the IP address to bind to. By default, it will bind to 0.0.0.0, meaning all available IP addresses.
- `amaflags`: Just as for SIP and DAHDI interfaces, we can set `amaflags` for IAX.
- `accountcode`: As for SIP and DAHDI interfaces, we can set an account code to be used for billing purposes.
- `language`: We can specify a language. If it is omitted, Asterisk will default to English.
- `bandwidth`: We can specify a level of bandwidth. This will automatically set which codecs can and cannot be used. The valid choices are "low", "medium", and "high".
- Codec selection: Just as in `sip.conf`, we can disallow all codecs, and only allow the codecs that we want to permit. If we use `allow=all`, it is the same as using `bandwidth=high`.
- `jitterbuffer`: This allows us to configure our jitter buffer. We should leave it at `jitterbuffer=no` unless our network is unusually jittery because of the added latency that using buffers will typically create.
- `trunkfreq`: We can set the number of milliseconds between trunk messages. The default of 20 ms should be good enough for most installations.
- `register`: Just as in `sip.conf`, we can register with a remote server. If our username is "johndoe", and our password is "foo", and we want to register with "reallycoolhost.com", then the statement would be: `register=>johndoe:foo@reallycoolhost.com`.
- `tos`: Here we set our TOS bits. Valid choices are "lowdelay", "throughput", "reliability", "mincost", and "none". Usually "lowdelay" is desirable.

Now that we have set our options in the general section, we need to define users. We can see that it is very similar to the way we configured SIP. One difference between the SIP and IAX protocols is that IAX will support RSA public/private key encryption, along with the MD5 and plaintext authentication methods. If we choose to use RSA authentication, then we must put the keys in `/var/lib/asterisk/keys/`. All public keys must end with `.pub`, while private keys will end with `.key`.

Configuring Asterisk

There are a few options that can be set per user entry, in addition to the settings already mentioned. They include:

- `type`: Just as in SIP, we can have "user", "peer", or "friend".
- `auth`: Can be "md5", "plaintext", or "rsa".
- `inkeys`: Name of the key file or files to use, with a colon between each acceptable public key for RSA-encrypted authentication.
- `outkey`: Name of the private key file to use for RSA-encrypted authentication.
- `secret`: The password to use.
- `notransfer`: If set to "yes", this will disable IAX's ability to use native transfers.
- Security: Much like the allow/disallow pair for codec selection, we can use permit/deny directives for IAX connections.
    - `deny`: List of all IP addresses and blocks to disallow access from. If we enter "0.0.0.0/0.0.0.0", then all access will be blocked.
    - `permit`: List of all IP addresses (as individual addresses or blocks) to permit access from. If we enter "192.168.1.100/255.255.255.255", then the user at 192.168.1.100 will be permitted access.
- `qualify`: We set this to "yes" when we want to make sure a user's device is connected before attempting to send calls to it.
- `trunk`: We set this to "yes" to use IAX trunking. Trunking refers to putting multiple phone calls in the same stream to save the overhead of packaging each conversation individually. If we will often have multiple conversations in progress with the same host (that is another Asterisk server uses this account), then it is a good idea to set `trunk=yes`.

Now we have enough information to create all of our IAX users, so we will go through an example. Suppose we have a user who will be on extension `2000`, with a password of `2000iscool`. This is a hardware IAX phone, and as such, will be dynamic and of the `friend` type. The user will be permitted to make long-distance telephone calls, and their telephone will send the correct caller ID name and number, and we will trust it. This user is only permitted to connect from our local subnet, which is the class C address of 192.168.1.0.

```
[2000]
type=friend
secret=2000iscool
host=dynamic
defaultip=192.168.1.200
```

```
callerid=asreceived
context=longdistance
deny=0.0.0.0/0.0.0.0
permit=192.168.1.1/255.255.255.0
```

We can now go through our listed devices and make similar entries for any of our terminal devices of the type of IAX.

Voicemail

Asterisk provides a voicemail program called Asterisk Mail. Using `etc/asterisk/voicemail.conf`, we can configure global options for our voicemail system as well as define different voicemail boxes.

In the configuration file, we first have our general options:

```
[general]
```

The first setting we get to decide on is how to write the voice files as they are recorded. The default is usually fine:

```
format=wav49|gsm|wav
```

Formats that don't match the codec of the call will require transcoding and all applicable licenses. For a system designer looking to minimize transcoding, making sure the voicemail will record in the applicable formats that are allowed for the calls will be beneficial.

Next, we get to set up email notification. Asterisk mail will allow us to notify users of new messages through email, and optionally, we can attach the voice files directly to the message. This is the reason we select `wav49` as the format in the previous code, as most computers will be able to play the files. We choose the email return address and whether we will attach the voicemail to the email with `serveremail` and `attach` respectively. We can also choose the display name that the email comes from, by setting `fromstring`. If we need to, we can also override the email sending program, in the `mailcmd` variable.

```
serveremail=asterisk@mydomain.tld
fromstring=Asterisk PBX
attach=yes
mailcmd=/usr/bin/sendmail -t
```

Now we need to set some limits on messages. We will define `maxmessage`, which is the maximum length, in seconds, of a voicemail message. The variable `minmessage` is similar, in that it is the minimum length, in seconds, of a voicemail message. We also set `maxgreet`, which is the maximum length our users can record for a greeting, again measured in seconds. Through `maxsilence`, we set how many seconds of silence we accept before ending the recording. Thus, we could set:

```
maxmessage=180
minmessage=3
maxgreet=45
maxsilence=5
```

These are not the only variables for controlling the general behavior of Asterisk Mail. We also have:

- `skipms`: This defines the number of milliseconds to move ahead or back when fast forwarding and rewinding the message.
- `maxlogins`: This sets how many failed attempts users can have in one session before being disconnected.
- `silencethreshold`: This allows us to configure what the system will consider to be silence. For the silence threshold, the smaller the number, the lower ambient noise will have to be before it is recognized as silence.

To continue our example, we could have:

```
skipms=2000
maxlogins=3
silencethreshold=128
```

There are more options available for us to change the body of the email, notify external programs of new voicemail messages, change our character set, and a whole host of other options. However, the default settings should work in most installations. If they do not meet our needs, we can configure them at any time.

Now that we have configured the general options, we have the option of creating time zone messages. This is how we let Asterisk Mail know what it should say when telling users about when messages arrived. For instance, if we lived in the Central time zone, we would have:

```
[zonemessages]
central=America/Chicago|'vm-received' Q 'digits/at' IMp
```

Each time zone is defined in `/usr/share/zoneinfo/`. Just find a city in the desired time zone to define the zone message.

The following is a larger example of zone messages:

```
[zonemessages]
madrid = Europe/Paris|'vm-received' Q 'digits/at' R
paris = Europe/Paris|'vm-received' Q 'digits/at' R
sthlm = Europe/Stockholm|'vm-recieved' Q 'digits/at' R
europa = Europe/Berlin|'vm-received' Q 'digits/at' kM
italia = Europe/Rome|'vm-received' Q 'digit/at' HMP
eastern = America/New_York|'vm-received' Q 'digits/at' Imp
central = America/Chicago|'vm-received' Q 'digits/at' Imp
pacific = America/Los_Angeles|'vm-received' Q 'digits/at' IMP
```

Finally, we have to define actual mailboxes. We have to place each mailbox in a context. This context should be the same as the context that the user's extension will appear in, and if possible, the voicemail box should be the same as the extension of the user. The format for the configuration line is: `voicemailbox => password,username,emailaddress,pageraddress,options`.

We can use one or more options for each user. Some of the options to be aware of are:

- `tz`: This sets the time zone, as defined in `[zonemessages]` in the previous code
- `attach`: This tells Asterisk Mail whether this particular user wants to have their voicemail file attached to the email
- `saycid`: This can make Asterisk Mail say the caller ID of the caller who left the message
- `operator`: This allows us to define whether the caller can press 0 to get an operator while leaving a message

Each option is separated by the pipe (|) symbol.

So, if we wanted to have a normal voicemail box for extension `1000`, with a password of `1234`, and accepting all of the defaults, and not sending email, we would have the following line in `voicemail.conf`:

```
1000 => 1234,Example Mailbox
```

Now, suppose we wanted to set up a little more of an advanced example. Suppose `Joe User` on extension `1001` with a password of `123456789` has an email address of `juser@domain.tld`, no pager, and lives in the Central time zone. He wants his voicemail message attached to his email, wants Asterisk Mail to speak the caller ID, and wants his callers to be able to press 0 for an operator. His voicemail box line would look like the following:

```
1001 => 123456789,Joe User,juser@domain.tld,,tz=central|attach=yes|
saycid=yes|operator=yes
```

Great! We have just configured a fully-functional voicemail box. In a similar fashion, we can now create all of the voicemail boxes we need.

Music on hold

Using music on hold, Asterisk enables us to stream MP3 or WAV files to any handset or line. These streams are commonly used for music on hold and for the music played while people are waiting in a queue. Each stream is configured in `/etc/asterisk/musiconhold.conf`.

Asterisk gives us the flexibility of defining multiple instances of MOH, referred to as classes. Each class can use a different directory of audio files and a different mode. For our purposes, we will only be using the mode called **files**. This mode allows us to use Asterisk's native players to stream music on hold.

Using Asterisk's native player also allows us to have our music on hold in various formats that Asterisk supports. Our PBX will determine the best format to play. Therefore, transcoding may be avoided if we create our music-on-hold files in the format our channels are in.

There are two main directives we need to be aware of:

- **Mode**: This refers to the mode we previously discussed. We will be using files, meaning we will use Asterisk's native player.
- **Directory**: This simply points Asterisk to the directory holding the music-on-hold files.

If we choose to shuffle the music-on-hold media files, we can simply add the directive called `random` and set it to `yes`.

A typical music-on-hold class would look like the following:

```
[default]
mode=files
directory=/var/lib/asterisk/music-on-hold
random=yes
```

That's all there is to it! We have now defined two different classes of music. Now, we need to put in an entry for each different stream of audio files we want our phone system to have.

Please note that using MP3 files can add an extra burden to your systems resources (CPU, and so on). For this reason many users today are using WAV files, and some even use GSM. A great utility to convert your audio files to WAV or GSM is "SoX". SoX is a sound file format converter found on most Unix systems. If your trying to convert your MP3 to WAV you will need to use mpg123 or lame.

Converting MP3 to WAV:

    mpg123 -w example.wav example.mp3

Converting Wav to GSM:

    sox example.wav -r 8000 -c1 example.gsm resample -ql

Queues

As we discussed in Chapter 2, queues give us a logical place to put callers until we are ready to answer the calls. Queues are a very flexible and powerful tool to improve customer service, and better utilize our personnel.

When we edit /etc/asterisk/queues.conf, the first sections we will have are [general] and [default]. Neither of them is used yet.

Below these headings, we come to where we will define our queues. For simplicity's sake, I recommend we name the queue according to the main extension that will represent it. For instance, if on our worksheet we entered that the queue would be dialed from extension 1000, I would define the queue as follows:

    [1000]

Now we need to set the parameters for the queue. First, we set the music that the callers will hear until their call is answered. This is done by setting the music variable as follows:

    music = q1000

We are stating that the queue should play the music indicated in the q1000 class of MOH, as defined in musiconhold.conf. Each queue can have its own music, or it can use the default MOH.

Next, we define an announcement, if we wish to have one. This will play an audio file to the agent when he or she answers the queued call. We might find this helpful if we have agents answering multiple queues, to tell them which queue the calls come from as they answer the calls. If we use this, be sure that we have a file in /var/lib/asterisk/sounds/<promptfilename>.<extension> that works:

    announce = <promptfilename>

Notice that we do not need to use the extension in the configuration file as Asterisk will recognize valid files. However, even though we don't have to put the extension in the filename in the configuration file, it must still have the correct extension in the /var/lib/asterisk/sounds directory.

Now we need to define our strategy. We should have decided this on our worksheet. For example, if we wanted to use the ringall strategy, we would enter:

```
strategy = ringall
```

If we do not define a strategy, Asterisk will default to ringall.

The following are different ring types one can use for a queue:

- ringall: Rings all available agents until one answers
- roundrobin: Take turns ringing each available agent
- leastrecent: Rings the agent who was least recently called by this queue
- fewestcalls: Rings the agent with fewest completed calls from this queue
- random: Rings a random agent
- rrmemory (round robin with memory): Remembers where we left off the last ring pass

We may now define an escape context. We have not yet explored the full meaning of a context, but will be doing so soon. If we checked the box indicating that our users should be able to escape, we will call the context by the name of this queue, followed by the word out. Therefore, for this example, it would be:

```
context = 1000out
```

Now we get to set the timeout. This is how long each handset will ring before the queue will consider the call unanswered. We should set this to be the longest reasonable time that we expect our agents to take to pick up the phone. In our case, our agents do not use headsets, but answer questions on the computer. Therefore, they should be given 15 seconds to answer the phone before we assume that they are away from their desk.

```
timeout = 15
```

Next, we define the amount of time we wait before trying all of the extensions again. As our company handles many calls, and we want our customer to have the shortest possible wait time, we will define it to be 0 seconds.

```
retry = 0
```

Now we can set a limit on how many calls will be enqueued. As we want to allow every caller to enter the queue, we will set this to `0`.

    maxlen = 0

On our worksheet, we decided if we wanted to announce callers' positions in the queue, along with an estimate of how long they can expect to be kept in the queue. First, we enter how often we want these announcements. As I want it to happen every two minutes, I will set it to `120` seconds. If I wanted to have no announcements, I would enter `0`.

    announce-frequency = 120

Now we set whether or not to give estimates of waiting time when the current queue position is announced. Valid options are `yes`, `no`, and `once`, meaning the queue application will announce the position only once, when the caller first enters the queue. As I do want to announce their estimated wait time, I enter:

    announce-holdtime = yes

Next, we can define filenames for the announcement. As the defaults work for us, we will not change them. If we started serving people who spoke a different language, we could change the recordings to use each user's native tongue.

Finally, we define our members. We have a bit of flexibility in defining this. We can either define them as being agents in the queue, or we can hardcode the members to be handsets. As we want to do things right, we will define the members to be agents.

    member => Agent/007
    member => Agent/777

Notice that this only means that we have two members assigned to the queue, agent `007` and agent `777`, as defined in `agents.conf`.

We can now go on and define other queues, in exactly the same way as above. There is no built-in limit to the number of queues we can define. It depends only on the resources of the server.

Conference rooms

We now configure our conference rooms. These are defined and configured using the file `/etc/asterisk/meetme.conf`.

This file is among the simplest configuration files we will encounter. When we open the file we see something like the following:

```
;
; Configuration file for MeetMe simple conference rooms
; for Asterisk of course.
;
[rooms]
;
; Usage is conf => confno[,pin]
;conf => 1234
;conf => 2345,9938
```

As you can see, we have only to define the conference number. We have the option of creating a PIN, to give some level of security to the conference room. In the previous configuration file, two conference rooms were created. The first had no password set, and is conference number `1234`. The second, conference `2345`, will require that people enter the password of `9938` before they can join the conference.

For ease of administration, I suggest we start off using the same conference number as the extension that we're going to call it. As we move around the configuration files we see the advantage of using an extension number as name, especially when trying to troubleshoot a configuration.

Summary

This chapter has walked us through the process of configuring the key parts of Asterisk needed by most PBX setups. We looked at settings for:

- `chan_dahdi.conf` and DAHDI (`system.conf`)
- Session Initiation Protocol, SIP (`sip.conf`)
- Asterisk's homebrew protocol, Inter-Asterisk eXchange, IAX (`iax.conf`)
- Voicemail (`voicemail.conf`)
- Music on hold (`musiconhold.conf`)
- Call queues (`queues.conf`)
- Conference rooms (`meetme.conf`)

The next step, as we discover in the next chapter, is to create a **dialplan** to tell Asterisk how to handle any particular incoming call.

5
Creating a Dialplan

Congratulations! We have now installed Asterisk and configured most of your new open source phone system. Some readers may now be asking, "Are we there yet?". Well, quit whining—it's not far now and we have already come a good way. Let's pause for a moment and review the plan that we created in Chapter 2, *Making a Plan for Deployment*, where we should have decided what extensions and services we'll provide. Relax, the hard part is behind us, and now we might even start having fun!

In this chapter, we are going to create a full dialplan together. What is a dialplan? Asterisk is a powerful and flexible solution for many different telephony needs. Did you ever wonder how we use all of these features? This is how. At this point, programming experience may well give you an advantage, but it is by no means essential.

When calls come into the switch, we tell Asterisk step-by-step how to handle the call. Steps can be as simple as playing a sound file to running a customized script. We are limited mostly by our imaginations at this point.

We define all the steps we want Asterisk to perform in our `extensions.conf` file, in the customary location `/etc/asterisk`.

Before we begin, we need to set `priorityjumping=yes` in the `[general]` section of `extensions.conf`. This will allow the tips and tricks in this chapter to work with Asterisk 1.6.x.

Creating a context

What is a context? Simply said, a context is a group of extensions. Each extension must exist within a context. There is more to contexts than grouping extensions though.

Creating a Dialplan

In our `extensions.conf` file (or any included files), a context is denoted by square brackets "[ ]", as shown:

```
[mycontext]
. . .
```

So, if a context is a group of extensions, why do we need more than one? Let's think for a minute. Not all employees should be able to dial every phone. Would you trust your 16-year-old intern with the ability to dial international calls? I wouldn't. Also, do you want your president to be bothered by customers in the waiting room who use a courtesy phone and misdial? We could find that hazardous to our continued employment.

Certain extensions are hidden or made inaccesible from other extensions by context. This gives us some level of security. It also allows us to host multiple phone systems on a single server.

Imagine you have two businesses on the same phone system, each with only two handsets. It'd be a pain to have each dial four digits to reach the other handset. We can use contexts to treat each company as if it were on a separate server.

Something very important about contexts is we can include other contexts through the use of the `include` directive. This means all extensions in an included context are available. The value of this may not be immediately apparent, but soon we will see the full power of this tool.

Suppose we have some context named `bob`. If we wanted `bob` to include `default`, then we would have the following in our `extensions.conf`:

```
[bob]
include => default
```

This single line placed in any context gives that context the ability to dial any extension in the default context, as well as all contexts included in the default context. This means that if the default context included the `foo` context, then anybody in the `bob` context could dial extensions in the `foo` context.

Suppose we had the following in our `extensions.conf` file:

```
[foo]
exten => 1,1,Playback(tt-monkeys)
include => bar

[bar]
exten => 1,1,Playback(tt-weasels)
```

Now I know that we haven't yet discussed the definition of extensions. That's OK. All we need to know is that extension *1* in `foo` will play back a file that sounds like monkeys, and extension *1* in `bar` will play back a file that says, "weasels have taken over our phone system".

If we are in context `foo` and press *1*, which file will play? This shows us the danger of `include`. We should be careful not to include multiple matches for the same extension. If we do include multiple contexts, the first included context with a match will win. Consider the following file:

```
[foobar1]
include => foo
include => bar

[foobar2]
include => bar
include => foo
```

If we are in context `foobar1` and press *1*, we will hear monkeys, while if we are in context `foobar2` and press *1*, we will hear weasels. While this can trip the unwary, we will use it to our advantage later on.

Creating an extension

We all have a good idea about what an extension is. On our legacy PBX, each handset was an extension. Pretty simple, right?

While conceptually simple, there is a little wrinkle. If all we want to do is provide a few handsets, then there's one extension per phone. But Asterisk can do much more! We need to think of an extension as a group of commands that tells Asterisk to do some things. As amorphous as that may be, it's true.

An extension can be tied to one handset, a queue, groups of handsets, or voicemail. An extension can be attributed to many different areas of the system. If you're familiar with programming terms, perhaps you could say that extensions are polymorphic.

To go further, extensions can be used to provide access to other applications, sound files, or other services of Asterisk. Extensions are important to the magic of Asterisk.

Now that we know why we create extensions, let's think about how we create them. Again, they are in the `extensions.conf` file, or any files that you include from there.

Creating a Dialplan

> We may decide to break up files such as `extensions.conf` into multiple configuration files. A common example of this is when we create large groups of extensions and choose to give each its own file. This also applies to the other configuration files we use.

The general format for a line in the `extensions.conf` file is:

```
exten => extensionnum,priority,action
```

Let's take a closer look. Each line begins with the command `exten`. This is a directive inside Asterisk. You do **not** change this for each extension.

Next, we have the extension number. Each extension has a unique number. This number is how Asterisk knows which set of commands to run. This extension can be detected in three major ways. First, the phone company may send it in with the calls, as is the case with DID numbers. Users can enter an extension using their touch-tone keys. Finally, there are a few special extensions defined. Some of these are:

- `s`: start extension. If no other extension number is entered, then this is the extension to execute.
- `t`: timeout extension. If a user is required to give input, but does not do so quickly enough, this is the extension that will be executed.
- `i`: invalid extension. If a user enters an extension that is not valid, this is the extension that will be executed.
- `fax`: fax calls. If Asterisk detects a fax, the call will be rerouted to this extension.

Then we have the priority. Asterisk will start at priority 1 by default, complete the requested command, and then proceed to priority *n*+1. Some commands can force Asterisk to jump to priority *n*+101, allowing us to route based on decisions, such as if the phone is busy.

Finally, we have the action. This is where we tell Asterisk what we want to do. Some of the more common actions we may want to perform are:

- `Answer`: This accepts the call. Many applications require that the call be answered before they can run as expected.
- `Playback(filename)`: This command plays a file in `.wav` or `.gsm` format. It is important to note that the call must be answered before playing.
- `Background(filename)`: This command is like `Playback`, except that it listens for input from the user. It too requires that the call be answered first.

- `Goto(context,extension,priority)`: Here, we send the call to the specified context, extension, and priority. While useful, this can be a bad style, as it can be very confusing to us if something goes wrong. However, it can be a good style if it keeps us from duplicating extension definitions, as moves, adds, or changes would only have to be updated in one place.
- `Queue(queuename|options)`: This command does what it seems like it should. It places the current call in the queue, which we should have already defined in the `queues.conf` file.
- `Voicemail(extension)`: This transfers the current call to the voicemail application. There are some special options as well. If we preceed the extension with the letter s, it skips the greeting. When we place the letter u before the extension, it uses the unavailable greeting, and b uses the busy greeting.
- `VoicemailMain`: This application allows users to listen to their messages, and also record their greetings and name, and set other configuration options.
- `Dial(technology/id,options,timeout)`: This is where we tell Asterisk to make the phone ring, and when the line is answered, to bridge the call. Common options include:
    - t: Allow the called user to transfer the call by pressing the # key.
    - T: Allow the calling user to transfer the call by pressing the # key.
    - r: Indicate ringing to the calling party.
    - m: Provide music on hold to the calling party.
    - H: Allow the calling party to hang up by pressing the * key.
    - g: Go on in the context if the destination hangs up.

While this list is not exhaustive, it should be enough to get us started. Suppose we just want to make a DAHDI phone ring, which is on interface 1, and we are going to work completely in the default context. Our `extensions.conf` file would look like:

```
[default]
exten => s,1,Dial(dahdi/1)
```

Pretty simple, right? Now, imagine we want to transfer to the voicemail of user 100 if someone is on the phone. As `Dial` sends you to priority *n+101* when the line is busy or not available, all we have to do is define what we want to do. Our dialplan would look like:

```
[default]
exten => s,1,Dial(dahdi/1)
exten => s,102,Voicemail(b100)
```

Creating a Dialplan

Great! We have some of the functionality that users have come to expect. But are you happy yet? The problem is that a phone could ring for years before someone picks it up.

So, for our next exercise, suppose we want to transfer the call to voicemail when the phone is not answered in 30 seconds. So, obviously, we're going to have to use the option in `Dial` to define a time-out. Our dialplan would have something like:

```
[default]
exten => s,1,Dial(dahdi/1|30)
exten => s,2,Voicemail(u100)
exten => s,102,Voicemail(b100)
```

All we're doing is telling Asterisk how to handle the call, in a step-by-step way. It is important to think about all scenarios that a call can go through, and plan for them. Just to reiterate a point I made earlier, planning ahead will save us hours of debugging later.

Suppose we want to send anyone who is in a place where they shouldn't be to user 0's voicemail, which will be checked periodically by the receptionist.

```
[default]
exten => s,1,Dial(dahdi/1|30)
exten => s,2,Voicemail(u100)
exten => s,102,Voicemail(b100)
exten => i,1,Voicemail(s0)
exten => t,1,Voicemail(s0)
```

All right, we're getting somewhere now! At least we know each call will be handled in some way. What about faxes? Suppose we have only one fax machine (or a centralized fax server) on DAHDI interface 2, then our dialplan should look similar to:

```
[default]
exten => s,1,Dial(dahdi/1|30)
exten => s,2,Voicemail(u100)
exten => s,102,Voicemail(b100)
exten => i,1,Voicemail(s0)
exten => t,1,Voicemail(s0)
exten => fax,1,Dial(dahdi/2)
```

Congratulations! We now have a working phone system. May be not the most interesting yet, but we're making great progress. Don't worry, our phone system will grow in features.

Now, to create a list of useful extensions, we need to define a set of commands for each handset we have. Suppose we have three SIP phone users—1001-1003, with extensions 1001-1003. Our default context would look like:

```
[default]
exten => 1001,1,Dial(SIP/1001|30)
exten => 1001,2,Voicemail(u1001)
exten => 1001,102,Voicemail(b1001)

exten => 1002,1,Dial(SIP/1002|30)
exten => 1002,2,Voicemail(u1002)
exten => 1002,102,Voicemail(b1002)

exten => 1003,1,Dial(SIP/1003|30)
exten => 1003,2,Voicemail(u1003)
exten => 1003,102,Voicemail(b1003)

exten => i,1,Voicemail(s0)
exten => t,1,Voicemail(s0)
exten => fax,1,Dial(dahdi/2)
```

For every extension we add, the length of `extensions.conf` will grow by four lines (three lines of code, and one line of whitespace). This is not very easy to read, and it is very easy to make mistakes. There has to be a better way, right? Of course there is!

We can use macros to define common actions. We will create a special **macro context**. The name of these contexts always starts with `macro-`. Suppose we want to call this one `macro-normal`. We would have:

```
[macro-normal]
exten => s,1,Dial(${ARG2}|30)
exten => s,2,Voicemail(u${ARG1})
exten => s,102,Voicemail(b${ARG1})
```

Now, to create the same three extensions, we would have:

```
exten => 1001,1,Macro(normal|1001|SIP/1001)
exten => 1002,1,Macro(normal|1002|SIP/1002)
exten => 1003,1,Macro(normal|1003|SIP/1003)
```

So now, each extension we add requires only one extra line in `extensions.conf`. This is much more efficient and less prone to errors. But what if we knew that any four-digit extension beginning with a 1 would be a normal, SIP extension?

Creating a Dialplan

Here it is time for us to discuss Asterisk's powerful pattern-matching capabilities. We can define extensions with certain special wildcards in them, and Asterisk will match any extension that fits the description.

Using the underscore (_) character warns Asterisk that the extension number will include pattern matching. When matching patterns, the X character represents any number (0 to 9), the Z character will match the numbers 1 to 9, the N character represents numbers 2 to 9, and the period (.) represents a string of any number of digits.

Also, we can use certain variables in our dialplan. One such variable is ${EXTEN}, which represents the extension that was used.

So, for this example, we could use the following definition:

```
exten => _1XXX,1,Macro(normal|${EXTEN}|SIP/${EXTEN})
```

This one line of code has now defined 1000 extensions, from 1000 to 1999. All we have to do is ensure that our voicemail user, extension, and SIP user are all the same number. Pretty cool, huh?

Note that if we wish to modify the behavior of all extensions, all we have to do is modify the macro. This should help us quite a bit as we tweak Asterisk to fit our business needs.

Creating outgoing extensions

With this dialplan, we have only catered for incoming calls. Of course we will want to create extensions to dial out.

How these outgoing extensions look depends on the plan we made earlier. It also depends on how you want the switch to act for your users. We always want to make it as similar to any old system as possible to reduce the need to retrain users.

Most phone systems require a user to dial a certain digit to designate the call as one that has a destination outside the switch. In our legacy PBX, we accessed outgoing lines by dialing a 9. To copy this behavior, we could do something like:

```
[outgoing]
exten => _9.,1,Dial(dahdi/g1/${EXTEN:1})
```

Notice that we are using pattern matching as we did before. Also notice that we used g1 as the DAHDI interface. This is a good use for a group. This simply means "any free interface in group 1". Thus, if we put all of our outgoing lines in group 1, when we dial an outside number, we do not have to guess which channel is free.

Remember that the variable EXTEN represents the extension that we are in. If a user dials *95555555*, then EXTEN is equal to *95555555*. By using ${EXTEN:1}, we instruct Asterisk to strip the first (leftmost) digit. Thus, ${EXTEN:1} equals *5555555*. If we wanted to strip the two leftmost digits, it would be ${EXTEN:2}.

Many discount long distance carriers will charge the same for local calls as long distance. In a case like that, we would want to make sure that local and toll-free calls went out on lines connected to the local telephone company, while calls destined for long distance locations should go out through our discount carrier. For this example, we will assume that the discount carrier is DAHDI group 1, and the local telephone company is DAHDI group 2.

```
[outgoing]
exten => _9N.,1,Dial(Zap/g2/${EXTEN:1})
exten => _91.,1,Dial(Zap/g1/${EXTEN:1})
```

Therefore, if the number dialed is preceded with a *9* and a *1*, then the call will go out through our DAHDI group 1 lines, and if it does not have a *1*, it will go out through our group 2 lines.

What if we had used the following?

```
[outgoing]
exten => _9X.,1,Dial(Zap/g2/${EXTEN:1})
exten => _91.,1,Dial(Zap/g1/${EXTEN:1})
```

Notice that this is a problem. When we dial *9*, followed by *1*, we have two statements that match. When you start Asterisk, you will notice that the system parses the extensions.conf (and included files) and immediately reorders all of the extensions before building a database file. Therefore, it is ambiguous as to which statement it will execute.

While this method is technically functional, it is not the best way to do it. A better way would be:

```
[outgoing]
exten => _9NXXXXXX,1,Dial(Zap/g2/${EXTEN:1})
exten => _91NXXNXXXXXX,1,Dial(Zap/g1/${EXTEN:1})
```

This does not allow for medium distance dialing. It also does not handle the case of all lines being busy in one of the groups. Let's see how we could take care of that:

```
[outgoing]
exten => _9NXXXXXX,1,Dial(Zap/g2/${EXTEN:1})
exten => _9NXXXXXX,2,Dial(Zap/g1/${EXTEN:1})
exten => _91NXXNXXXXXX,1,Dial(Zap/g1/${EXTEN:1})
exten => _91NXXNXXXXXX,2,Dial(Zap/g2/${EXTEN:1})
exten => _91NXXXXXX,1,Dial(Zap/g1/${EXTEN:1})
exten => _91NXXXXXX,2,Dial(Zap/g2/${EXTEN:1})
```

This is getting much better, but what when all of our lines are busy for both groups? We should probably notify the user that their call didn't go through because all lines are in use. We could do something like this:

```
[outgoing]
exten => _9NXXXXXX,1,Dial(Zap/g2/${EXTEN:1})
exten => _9NXXXXXX,2,Dial(Zap/g1/${EXTEN:1})
exten => _9NXXXXXX,3,Congestion
exten => _91NXXNXXXXXX,1,Dial(Zap/g1/${EXTEN:1})
exten => _91NXXNXXXXXX,2,Dial(Zap/g2/${EXTEN:1})
exten => _91NXXNXXXXXX,3,Congestion
exten => _91NXXXXXX,1,Dial(Zap/g1/${EXTEN:1})
exten => _91NXXXXXX,2,Dial(Zap/g2/${EXTEN:1})
exten => _91NXXXXXX,3,Congestion
```

Imagine a phone switch, in which all extensions were exactly four digits long. How would this knowledge affect our outgoing lines? Perhaps the fundamental question is whether we really need to dial the 9 at all. While people generally expect to dial a 9 at work, they do not seem to care to do so from home. May be we should do this:

```
[outgoing]
exten => _NXXXXXX,1,Dial(Zap/g2/${EXTEN:1})
exten => _NXXXXXX,2,Dial(Zap/g1/${EXTEN:1})
exten => _NXXXXXX,3,Congestion
exten => _1NXXNXXXXXX,1,Dial(Zap/g1/${EXTEN:1})
exten => _1NXXNXXXXXX,2,Dial(Zap/g2/${EXTEN:1})
exten => _1NXXNXXXXXX,3,Congestion
exten => _1NXXXXXX,1,Dial(Zap/g1/${EXTEN:1})
exten => _1NXXXXXX,2,Dial(Zap/g2/${EXTEN:1})
exten => _1NXXXXXX,3,Congestion
```

There is a drawback to this configuration. If you are using DAHDI interfaces, and people are touch-tone dialing, Asterisk will have to wait for a timeout period to complete dialing some four-digit extensions. The reason for this is simple. Any extension number that partially matches the beginning of the patterns above will have to be held to see if more digits are coming. However, SIP and IAX phones do not seem to suffer the same issue.

Because of this issue, from here we will assume that all outgoing calls are preceded with the digit 9.

We have already discussed that contexts can provide security for outgoing phone calls, but this example fails to describe such security. Suppose you have two groups of employees, those who may make toll calls, and those who may not. All employees have an individual handset. The most logical choice is to make two contexts:

```
[longdistance]
exten => _9NXXXXXX,1,Dial(Zap/g2/${EXTEN:1})
exten => _9NXXXXXX,2,Dial(Zap/g1/${EXTEN:1})
exten => _9NXXXXXX,3,Congestion
exten => _91NXXNXXXXXX,1,Dial(Zap/g1/${EXTEN:1})
exten => _91NXXNXXXXXX,2,Dial(Zap/g2/${EXTEN:1})
exten => _91NXXNXXXXXX,3,Congestion
exten => _91NXXXXXX,1,Dial(Zap/g1/${EXTEN:1})
exten => _91NXXXXXX,2,Dial(Zap/g2/${EXTEN:1})
exten => _91NXXXXXX,3,Congestion


[local]
exten => _9NXXXXXX,1,Dial(Zap/g2/${EXTEN:1})
exten => _9NXXXXXX,2,Dial(Zap/g1/${EXTEN:1})
exten => _9NXXXXXX,3,Congestion
```

We simply place each handset into one of the two contexts, based upon what numbers we want them to be able to dial. While this would work, we duplicated three lines between the contexts. Remember when we discussed contexts, and I mentioned that we can use the `include` directive? Here's a good place to do so. So now, we have:

```
[longdistance]
exten => _91NXXNXXXXXX,1,Dial(Zap/g1/${EXTEN:1})
exten => _91NXXNXXXXXX,2,Dial(Zap/g2/${EXTEN:1})
exten => _91NXXNXXXXXX,3,Congestion
exten => _91NXXXXXX,1,Dial(Zap/g1/${EXTEN:1})
exten => _91NXXXXXX,2,Dial(Zap/g2/${EXTEN:1})
exten => _91NXXXXXX,3,Congestion
include => local
[local]
exten => _9NXXXXXX,1,Dial(Zap/g2/${EXTEN:1})
exten => _9NXXXXXX,2,Dial(Zap/g1/${EXTEN:1})
exten => _9NXXXXXX,3,Congestion
```

Let's take a quick look at what we have. We can now place handsets in the `local` or `longdistance` context. This gives us a good bit of security, but there is one small problem. Do you see it? If we put our users into either context, they cannot dial internal extensions. Assuming we allowed calling our local extensions in the default context, we should update the `local` context like so:

```
[local]
exten => _9NXXXXXX,1,Dial(Zap/g2/${EXTEN:1})
exten => _9NXXXXXX,2,Dial(Zap/g1/${EXTEN:1})
exten => _9NXXXXXX,3,Congestion
include => default
```

By so doing, we give access to the extensions for the `local` context. We also give access to the default context for those in the `longdistance` context because `longdistance` includes `local`, which in turn includes `default`.

Thus, we must be very careful about what we include. When we include any context, we are in turn including all contexts it includes. It is easy to include our way out of the security we set up.

Advanced Call Distribution

What exactly is Advanced Call Distribution? Many phone systems tout this feature, but most do not adequately define what it means. Basically, it refers to using call queues, parking calls for another user to answer, and Direct Inward Dialing (DID).

So that we keep our focus, we will look at each of these elements individually.

Call queues

In Chapter 4, *Configuring Asterisk*, we configured call queues through the `/etc/asterisk/queues.conf` file. As we go through how we're going to use our queues, we may decide we want to change the way our queues are configured. There is absolutely no problem with changing the configuration so that it more accurately reflects our needs. Just remember that we need to issue a reload on the Asterisk console, or type `#asterisk -r -x reload` at the command line.

The power and flexibility of other ACD systems can be matched or exceeded by Asterisk. As we evaluate our needs, we should remember that configuring a single aspect of Asterisk sometimes requires changes to more than one file. For example, queues will be configured both in the `queues.conf` file and the `extensions.conf` file. As we have already discussed the `queues.conf` file in Chapter 4, we will discuss how to set up `extensions.conf` to give us the desired result.

When dealing with call queues, we need to think about the two types of users we have. First, we have the caller who calls in and waits in the queue for the next agent. We can think of this person as our customer. Next, we have the agents who work the queue. We can think of these people as our users.

As a business, we have to decide what we want our customers' experience to be. Our call queue can make it sound like a phone is ringing. Or we can use music on hold while the customer waits. We can also announce call position and estimated wait time if we want to.

When we place customers in a queue, we use the Queue application. To place a caller in the queue named bob, we would use something like:

    exten => 1000,1,Queue(bob)

Suppose we have an operator's extension. As Ollie the operator may have more than one call at a time, we decide to give him a call queue. His calls are always about a minute long. The customers waiting for him are going to be there because they got lost in a system of menus. His queue will be named operator.

In this instance, we will choose to have the customer hear the ring, so they will believe they are about to be helped. The sound of ringing should not last more than about a minute. We will not announce call queue length because our customer should not know that he or she is in a queue.

The entry for this queue would be:

    exten => 0,1,Queue(operator|tr)

Notice our use of options. Options for the queue application include:

- t: Allow the user to transfer the customer.
- T: Allow the customer to transfer the user.
- d: This is a data-quality call.
- H: Allow the customer to hang up by hitting *.
- n: Do not retry on timeout. The next step in the dialplan will be executed.
- r: Give the customer the ringing sound instead of music on hold.

Thus, we told the Queue application to make the customer hear the ring, and the user (Ollie) the ability to transfer calls (as he's the operator).

Creating a Dialplan

Now, suppose we have Rebecca, the receptionist at SIP phone `1006`. When Ollie goes to the bathroom, we want our poor lost customers to be routed to her. So we could use the following in our `extensions.conf` file:

```
exten => 0,1,Queue(operator|trn)
exten => 0,2,Dial(SIP/1006)
```

Now, Rebecca had better answer this. Until she does, the phone will continue to ring. Notice that this call will never end up in Rebecca's voicemail, as it is not transferred to her extension, but instead dials her phone directly.

We have adequately addressed the customer's experience. But now we need to look at how our users will join and leave the queue. Previously, we discussed the power and flexibility of using agents in queues. As with most things in Asterisk, there are many ways we can associate members to queues. The three main ways are—statically, dynamically, and by using agents.

Our first option is to have members statically assigned to the queue. In order to do this, we use the `member` directive in the `queues.conf` file. This is most helpful when we have a queue with fixed members, such as a switchboard queue.

Our second option is to allow members to log in dynamically. We do this through the `AddQueueMember` application. An example of this would be:

```
exten => 8101,1,AddQueueMember(myqueue|SIP/1001)
```

Whenever anybody dials extension `8101`, the telephone handset `SIP/1001` would be added to the queue named `myqueue`. All that we would have to do is define a login extension for every member of every queue.

What happens when this member no longer wishes to be in the queue? We use the `RemoveQueueMember` application, like this:

```
exten => 8201,1,RemoveQueueMember(myqueue|SIP/1001)
```

With this configuration, whenever anybody dials extension `8201`, the telephone handset at `SIP/1001` is removed. Again, we would have to define a logout extension for each member of the queue.

Suppose we did not wish to define a login and logout extension for each member. We have the option of leaving off the interface (`SIP/1001` in the previous example) and having Asterisk use our current extension. While this is very useful, Asterisk does not always use the right value. However, if it works for all extensions that need to be in the queue, we would only have to define one login and one logout per queue. The code would look like:

```
exten => 8101,1,AddQueueMember(myqueue)
exten => 8201,1,RemoveQueueMember(myqueue)
```

This is better than having to define a login and logout for each member of each queue, but sometimes users are not good at remembering multiple extensions to dial. The `AddQueueMember` application will jump to priority *n*+101 if that interface is already a member of the queue. Therefore, we could define an extension like:

```
exten => 8101,1,Answer
exten => 8101,2,AddQueueMember(myqueue)
exten => 8101,3,Playback(agent-loginok)
exten => 8101,4,Hangup
exten => 8101,103,RemoveQueueMember(myqueue)
exten => 8101,102,Playback(agent-loggedoff)
exten => 8101,105,Hangup
```

When we define it this way, a user dialing extension `8101` is logged in if not already a member of the queue, or logged out if in the queue. Also, we added a confirmation to the action, so that the user can know if they are now in or out of the queue. Notice that before we could use the `Playback` application, we had to answer the call. If we have a lot of these, we could define a macro extension, like:

```
[macro-queueloginout]
exten => s,1,Answer
exten => s,2,AddQueueMember(${ARG1})
exten => s,3,Playback(agent-loginok)
exten => s,4,Hangup
exten => s,103,RemoveQueueMember(${ARG1})
exten => s,104,Playback(agent-loggedoff)
exten => s,105,Hangup
...
[default]
exten => 8101,1,Macro(queueloginout|queue1)
exten => 8102,1,Macro(queueloginout|queue2)
exten => 8103,1,Macro(queueloginout|queue3)
```

And thus we see that using a macro will save us five lines in our `extensions.conf` for every queue after the first. This is how we can add queue members dynamically.

Our final option for adding queue members is by using Asterisk's agent settings. We were able to define agents in `/etc/asterisk/agents.conf`. We create an agent by defining an ID and a password, and listing the agent's name.

In the `queues.conf`, we could define agents as members of queues. Calls will not be sent to agents unless they are logged in. In this way, queues can be both dynamic and static—they are static when we do not change the members of the queues, but dynamic when calls will go to different handsets based upon which agents are logged in.

Creating a Dialplan

There are two main types of agents in this world. There are the archetypical large call center agents who work with a headset and never hear rings, and there are the lower-volume agents whose phone rings each time a call comes in. Asterisk has the flexibility to handle both types of agents, even in the same queue.

First, imagine a huge call center that takes millions of phone calls per day. Each agent is in multiple queues, and we have set each queue to use an announcement at the beginning of calls to let the agent know which queue the call is coming in from. As employees arrive for their shift, they sit down at an empty station, plug in their headset, and log in. Each employee will hear music in between calls, and then hear a beep, and the call will be connected. To accomplish this, we use the line:

```
exten => 8001,1,AgentLogin
```

Through the normal login, the call is kept active the whole time. The agents will logout by hanging up the phone. This allows large call centers to be quieter, as the distraction of ringing phones will be removed. It also allows for more efficient answering of lines, as the time required to pick up the phone is eliminated.

When our users arrive at work and wish to log in, they call extension `8001`, where they are prompted for their agent ID, password, and then an extension number at which they will take calls. This is how Asterisk knows how to reach them. Our agents can log out when using `AgentCallbackLogin` by going through the same procedure as for login, with the exception that when they are prompted for their extension, they press the # key.

It may be a good idea for us to review `agents.conf`. If we defined `autologoff`, then after the specified number of seconds of ringing, the agent will be automatically logged off. If we set `ackcall` to `yes`, then agents must press the # key to accept calls. If we created a `wrapuptime` (defined in milliseconds), then Asterisk will wait that many milliseconds before sending another call to the agent. These options can help us make our phone system as user friendly as we want it to be.

Through the use of call queues, we can distribute our incoming calls efficiently and effectively. We have plenty of options, and can mix and match these three ways of joining users to queues.

Call parking

In many businesses across the United States, an operator can be heard announcing "John, you have a call on line 3. John, line 3." In Asterisk, we don't really have lines the way analog PBXs do. Our users are accustomed to not having to transfer calls, especially when they may not know exactly where John is.

Asterisk uses a feature known as call parking to accomplish this same goal. Our users will transfer calls to a special extension, which will then tell them what extension to call in order to retrieve the call. Then our users can direct the intended recipient to dial that extension and connect to the call.

In order to be able to use this feature, we must define our parking lot. This is done in the `/etc/asterisk/parking.conf` file. In this file, there are only a few options that we will need to configure. First, we must create the extension that people are to dial in order to park calls. This can be whatever extension is convenient for us. Then we will define a list of extensions on which to place parked calls. These extensions will be what users dial to retrieve a parked call. Next, we will define what context we want our parked calls to be in. Finally, we will define how many seconds a call remains parked before ringing back to the user who parked it. Here is an example:

```
[general]
parkext => 8100
parkpos => 8101-8199
context => parkedcalls
parkingtime => 120
```

These settings would mean that we can park calls by dialing `8100`, and the call will be placed in extensions `8101` through `8199`, giving us the ability to have up to 99 parked calls at any given time. The calls will be in the context called `parkedcalls`, which means we should be careful to include it in any context where users should be able to park and retrieve calls.

When our users transfer a call to extension `8100`, they will hear Asterisk read out the extension that the call has been placed on. They can now make a note of it and notify the appropriate co-worker of the extension to reach the calling customer on. If the call is not picked up within the given `parkingtime`, then the call will ring back to the user who parked the call.

By using call parking, we can help our users by providing a feature similar to that of previous generations of PBXs. This also allows users to collaborate and redirect callers to other users who are better equipped to handle our customers' needs.

Direct Inward Dialing (DID)

Suppose we work at a healthcare company with over 100 employees. We have two PRI lines coming in, and only three switchboard agents to handle incoming calls. As a healthcare company, we schedule many appointments, answer questions about prescriptions, and help patients with billing questions. These three agents are always busy.

Creating a Dialplan

Now suppose the IT guy's wife calls in to ask if he wants sprouts or mash with his dinner. Do we want our switchboard agents to have to answer the call, find out who it is and what they want, and then transfer the call, or would we rather want the IT guy's wife to call her husband directly?

This is where **Direct Inward Dialing (DID)** comes in handy. DID is a service provided by phone companies where they send an agreed-upon set of digits, depending on the number the customer dialed. For most phone companies, the sent digits will be the full ten-digit number (in the United States). But this can be as small as the last digit.

All right, so the phone company is sending digits. What are we going to do with them? Imagine you have a PRI coming in to your office, and only ten phone numbers—a block from (850) 555-5550 to 5559. Your phone company has agreed to send you only the last digit dialed, which will be from 0 to 9, because you are guaranteed for this to be unique. Asterisk can route calls based on this DID information.

If we have our PRI line's channels defined to go into a context called `incoming`, this context could look like:

```
[incoming]
s,1,Goto(default,s,1)
i,1,Goto(default,s,1)
t,1,Goto(default,s,1)

0,1,Goto(default,1234,1)
1,1,Goto(default,2345,1)
2,1,Goto(default,3456,1)
3,1,Goto(default,4567,1)
4,1,Goto(default,5678,1)
5,1,Goto(default,6789,1)
6,1,Goto(default,7890,1)
7,1,Goto(default,1111,1)
8,1,Goto(default,1111,1)
9,1,Goto(default,1111,1)
```

There are a few things we should notice about this. First, we handled the error cases. What if a glitch at the phone company results in four digits being sent? We cannot allow a simple mistake on their end to interrupt our ability to receive phone calls.

Secondly, we are using `Goto` statements. We've briefly discussed how they can be both good and bad. In this case, if a user moves from one extension to another by using `Goto`, we have to update it only in the default context.

Finally, we are allowed to send multiple incoming DIDs to the same extension, if we so desire, as in the last three lines shown in the previous code. This might be useful if extension `1111` is the operator, and we do not yet have the number `7`, `8`, or `9` assigned to a user.

Of course, in real life this is going to get much more complicated, as phone numbers will probably come in with the full ten digits. But the concept is the same—we can define extensions based upon information that the phone company sends when the call is established.

By using DIDs, we can cut down on bottlenecks and give direct access to certain extensions. This tool of Asterisk helps make our phone system fast, efficient, and friendly to our users and customers.

Automated attendants

Any time we call a large company, we are greeted by a computer voice, asking us to route our call based on what we want or need. We are all familiar with call menus. While we won't get into a philosophical debate about how good or bad they are, we will talk about how to make them.

Suppose we want to create a menu of options such as, "For a billing question, press *1*, to request a configuration change press *2* ….". Now suppose you press *1*, and you hear the option, "For help reading your statement, press *1*, if you wish to dispute a charge, press *2*, …". This is just a standard phone tree, with which most users are comfortable. Asterisk knows which extension to execute based upon what context we are currently in.

Suppose that your customer service representatives are on SIP/1000, and the manager whom you wish to handle all disputes is on SIP/1001. Then, you have technicians on SIP/1002 and SIP/1003. Our configuration file could look like:

```
[mainmenu]
exten => s,1,Answer
exten => s,2,DigitTimeout(5)
exten => s,3,ResponseTimeout(30)
exten => s,4,Background(welcome)
exten => s,5,Background(options)

exten => 1,1,Goto(billing,s,1)
exten => 2,1,Dial(SIP/1002&SIP/1003)

[billing]
exten => s,1,Answer
exten => s,2,DigitTimeout(5)
```

```
exten => s,3,ResponseTimeout(30)
exten => s,4,Background(billingoptions)

exten => 1,1,Dial(SIP/1000)
exten => 2,1,Dial(SIP/1001)
```

There are some very important points to make here. It is good to define your digit and response timeouts each time you are going to give an option to the user. This makes your dialplan easier to understand and maintain. Also, notice that we must use the `Answer` command before being able to play back files. Remember that `Background` allows for user input to be captured during the file, while `Playback` does not.

Also notice that we issued the `Answer` command in the billing menu as well. If we knew that nobody could ever get into the billing menu without having passed through the main menu, we could probably leave the command out. However, it does not have any serious adverse consequences, and by leaving it in, we are able to offer a direct line to the billing department, if we ever choose to do so.

Suppose we want to have users to be able to dial any extension at any time. All we have to do is use the magical `include => default` in each of the contexts. This can cause the same delay in dialing that we discussed previously in the section *Creating outgoing extensions*, so it should be used with care. Also, users sometimes mash buttons at random, and may stumble across random extensions, frustrating the user. Some organizations have chosen to have users press a specific digit to be able to dial extensions directly to deal with these problems.

On another design note, most customers do not like being trapped for long periods of time in menu systems. Care should be taken to ensure menus do not get too deep. Also, it is easier for customers if you don't give too many options in one level.

Care must also be taken in selecting invalid and timeout responses. But most importantly, we have to do something. Leaving a customer in limbo with nowhere to go and no prompts to get them there is not a friendly move.

Let's build on our previous system with these concepts in mind. Of course, the prompts would be updated to suggest using the new features. For the purpose of this example, let's assume our receptionist is on SIP phone `1004`.

```
[mainmenu]
exten => s,1,Answer
exten => s,2,DigitTimeout(5)
exten => s,3,ResponseTimeout(30)
exten => s,4,Background(welcome)
exten => s,5,Background(options)
```

```
exten => 1,1,Goto(billing,s,1)
exten => 2,1,Dial(SIP/1002&SIP/1003)
exten => 3,1,Goto(dialbyext,s,1)

exten => t,1,Goto(s,1,1)
exten => i,1,Goto(s,1,1)

exten => 0,1,Dial(SIP/1004)

[billing]
exten => s,1,Answer
exten => s,2,DigitTimeout(5)
exten => s,3,ResponseTimeout(30)
exten => s,4,Background(billingoptions)

exten => t,1,Goto(billing,s,1)
exten => i,1,Goto(billing,s,1)

exten => 1,1,Dial(SIP/1000)
exten => 2,1,Dial(SIP/1001)
exten => *,1,Goto(mainmenu,s,1) ; escape to the previous menu

exten => 0,1,Dial(SIP/1004)

[dialbyext]
exten => s,1,Answer
exten => s,2,DigitTimeout(5)
exten => s,3,ResponseTimeout(30)
exten => s,4,Backgroud(enterextension)

exten => i,1,Playback(invalid)
exten => i,2,Goto(dialbyext,s,1)

exten => t,1,Playback(imsorryididntgetthat)
exten => t,2,Goto(dialbyext,s,1)

exten => *,1,Goto(mainmenu,s,1)

exten => 0,1,Dial(SIP/1004)

include => default
```

This is a pretty good system. We can build whatever we need using these concepts. Each menu system will be different, based on your needs. The menu system is easy to update according to your changing needs. As a reminder, when you have updated the dialplan, you can refresh it by executing the `reload` command at the console. This command does not interrupt calls that are currently in progress.

We should take a moment and discuss what a good menu is and what a bad menu is. We should put our most commonly chosen option first. We should not confuse our customers with too many choices. We should have errors that take you back one step, instead of all the way to the beginning of the menu. We should give an operator who can help people if they get lost. Finally, we should strive not to have our menu any deeper than four steps.

If we keep these design principles in mind, our automated attendant can be the best employee we've ever had. No demands, never sick, and always cheerful and ready to help!

System services

We have talked about how to create contexts, extensions, and how to make our system powerful. Some of the ideas we have discussed will be useful in some situations, yet not applicable in others. All of them work together to make Asterisk a flexible solution for many different needs. There are still other uses that we won't get into, as the need for them is less frequent.

There are some basic system services that we have not yet discussed. In Chapter 4, we configured our voicemail users. Asterisk also includes an application called the directory that reads the voicemail configuration and allows callers to look up an extension based on the user's last name. In the section *Creating an extension* earlier in this chapter, we saw how to send calls to voicemail. How do users retrieve messages? How do we use this directory?

```
exten => 8000,1,VoicemailMain(@default)
exten => 8888,1,Directory(default)
exten => 8888,2,Goto(1)
```

As you can see, providing access to the directory (extension `8888`) and voicemail (extension `8000`) is easy. But priority `2` of extension `8888` may seem odd. This is in here because the directory seems to crash from time to time. By setting priority 2 to place the user back into the directory, the failure is almost transparent to our users.

In priority 1 of extension 8000, we tell Asterisk to send people to voicemail in the default context. If we have only one context, we can usually get away without defining the context. However, it is better to be safe. In the directory application, we must define what context to use, which we specified as default. Both of these contexts should match the context listed in the voicemail.conf file.

When our users call the directory, they will be prompted to enter the first three letters of the person's last name. Pulling from the voicemail.conf file by default, Asterisk will search for all matches. If users have recorded their name (option 0, 3 in VoicemailMain), then Asterisk will play the file where the person speaks their name, otherwise, Asterisk will spell the complete name out. When it finds the person our user is looking for, they press 1. Asterisk immediately tries to dial that person.

Here we need to talk about naming our contexts. If our voicemail context is called foo, then Asterisk will try to dial extension@foo. However, if our extensions are all defined in a context called bar, then the directory will fail. Therefore, we must make sure that the contexts in voicemail.conf (where we define the voicemail entry) and the contexts in extensions.conf (where we define the extension) match.

Now suppose we want to record prompts for our menus. Asterisk can play standard Windows .wav files. However, getting the files recorded and into the phone system may not be the most convenient thing to do. Therefore, we can create a simple extension to allow us to record a prompt. We will allow users to input a four-digit name for the file so that they can record many prompts before having to sort them. The prompts will be stored in /tmp, and be recorded as .wav files. Let's assume we have files called enter4digits and record-instructions, as well as a file called 1toaccept2torerecord3torecordanother.

```
exten => 8200,1,Goto(record,s,1)

[record]
exten => s,1,Answer
exten => s,2,Read(RECORD|enter4digits|4)
exten => s,3,Playback(record-instructions)
exten => s,4,Record(/tmp/recording-${RECORD}|wav)
exten => s,5,Wait(2)
exten => s,6,Playback(/tmp/recording-${RECORD})
exten => s,7,ResponseTimeout(10)
exten => s,8,Background(1toaccept2torerecord3torecordanother)

exten => 1,1,Hangup
exten => 2,1,Goto(s,3)
exten => 3,1,Goto(s,2)
```

Creating a Dialplan

This little context will give us the option of recording whatever custom prompt we want, in a .wav format. We could add a timeout extension and an invalid extension if we so desire, just as we have done in other contexts.

Another great service that Asterisk provides is conferencing. We configured meetme.conf in Chapter 4 so that we could have conference rooms. For users to be able to enter the conference rooms, we must create an extension giving us access.

Suppose we have 10 conference rooms, which we want to place on extensions 8900 to 8909. We also named our conferences 8900 to 8909 in meetme.conf. Our extensions.conf should contain:

```
exten => 8900,1,MeetMe(8900)
exten => 8900,2,Goto(default,s,1)

exten => 8901,1,MeetMe(8901)
exten => 8901,2,Goto(default,s,1)

exten => 8902,1,MeetMe(8902)
exten => 8902,2,Goto(default,s,1)

. . .

exten => 8909,1,MeetMe(8909)
exten => 8909,2,Goto(default,s,1)
```

If we use a little bit of variable magic, we can get these lines down to:

```
exten => _890X,1,MeetMe(${EXTEN})
exten => _890X,2,Goto(default,s,1)
```

So, we see that by making our MeetMe conference number the same as the extension number that users dial to join, our lives are made a little bit easier. Also, by having all our conferences in a block of 10 or 100, we are able to use pattern matching to make our extensions.conf shorter.

Another service we may wish to have is a MusicOnHold extension. It will do nothing but play the music on hold that is currently running. This can be a useful tool for the administrator to check if the music is even running, or to check the volume on a handset. In order to add a MusicOnHold extension, we would add something like:

```
exten => 8010,1,MusicOnHold(default)
```

This extension will play the music that is playing on hold in the default class, as configured in the `musiconhold.conf` file. It will only end when the caller hangs up the phone. But we need to remember that it is not like the background music of some phone systems, in that it does tie up the line that is listening to the music. If calls come in while a user is listening to the music on hold, they will go through the normal procedure for a phone being busy.

Summary

In this chapter, we have looked at how to create a dialplan for our Asterisk system. We looked at how to set up:

- Contexts
- Extensions for incoming calls
- Extensions for outgoing calls
- Call queues
- Call parking
- Direct Inward Dialing
- Voicemail
- An automated phone directory
- Conference rooms

Although this list of available services is not exhaustive, it is certainly enough to get our phone system up and running. There are many more options available to us, which we will try out as we work through various case studies later in this book. These case studies will give us the full configuration of some Asterisk installations for fictitious companies. We will be able to view the configuration files and keep those parts that help us meet our particular needs.

6
Quality Assurance

The world has changed quite a bit in the last 150 years. Over this time, the telephone system has been invented, improved, and automated. Telephone switches no longer refer to people sitting in a large room connecting wires between the appropriate jacks. Flexible and powerful telephone service has moved from a dream to an expectation in large businesses, and for most of us it is a necessity.

Today, telephone systems are the lifeblood of business. They are how we take orders, acquire supplies, and even call for emergency assistance. With the increase in prominence of telephones, the expectations of telephone users have increased proportionally.

Not only have the technological expectations for telephone systems increased dramatically, but consumers are expecting more and more out of the businesses they call. Customers expect to be helped quickly and professionally. They want to know everything in a matter of minutes. Roads do not hold the only rage our society is facing today. As a business we have a variety of questions relating to our telephone system such as:

- How are our personnel handling angry callers?
- Are our employees answering the calls in a reasonable amount of time?
- Do we have any workers using the phone system for personal calls when they should be doing their job?

We will never be able to make sure everybody does what they are supposed to do all of the time. What we will be able to do at the end of this chapter is perform spot-checks on how we are doing on customer service, and make sure our phone service isn't being used for unauthorized purposes. Ultimately, it comes down to a matter of trust; however, some people do not know better because they haven't been fully trained. Most will always act honorably; however, some just cannot and should not be trusted. We will try to find out who is who.

Call Detail Records

When we talk about *security*, what images come to mind? May be a big, burly guard? Perhaps a bunch of guys in green, carrying machine guns? Do we imagine a person with a metal-detecting wand? Or do we think of thick glass window panes?

> All of these are security features. It is just that some are a little more intrusive than others. Each time we increase security, we become a little bit less friendly. We all have to decide how far we are willing (and able) to go.

In the continuum of security, Call Detail Records are the least intrusive. No special usernames or passwords have to be remembered. No fear of big brother breathing down your customers' and users' necks need be felt. We are simply doing the same thing telephone companies do—tracking what calls were made, when they were made, how long they lasted, where they came from, and a few other bits of information. This information is then available for us to review at our leisure.

Asterisk gives us a few options on how we track this information. The two major choices are flat-file logging and database logging.

Flat-file CDR logging

By default, Asterisk includes a module called `cdr_csv`. Right out of the box, Asterisk logs all calls coming in and going out. The information for these calls is placed in a **Comma Separated Value (CSV)** file. This CSV file is located in `var/log/asterisk/cdr-csv`. All information is available in `Master.csv`, and some channels can be configured to send some information to other files as well.

The benefit of using a CSV file is the simplicity. Right after compiling and installing Asterisk, this method will work. No additional configuration is required. Also, no additional network traffic is generated, and no additional services have to be installed on our server.

When using the CSV form of CDR, we will see lists and lists of values. They are not very easy to parse, so here is the format, in the order in which they appear:

- account code: As determined by the channel (for DAHDI) or the user (for IAX and SIP)
- source: The source of the call
- destination: The destination of the call
- destination context

- caller ID
- channel: The channel of the source
- destination channel: If applicable
- last application: The last application run on the channel
- last application argument: The last argument to the last application on the channel
- start time: The time the call commenced
- answer time: The time the call was answered
- end time: The time the call ended
- duration: The difference between start time and end time
- billable seconds: The difference between answer time and end time, which *must* be less than the duration
- disposition: Either ANSWERED, NO ANSWER, or BUSY
- `amaflags`: As set for the channel or user, like account code
- `uniqueid`: A unique call identifier
- `userfield`: A user field set by the `SetCDRUserField` command

We see that there are many items of information logged for each and every call. We can compare the billable seconds with our phone bill at the end of the month to make sure they're close. We can look at the destination and figure out if the calls were authorized. This gives us enough information to answer most questions we may have about a phone call.

While we have enough information to answer questions, finding that answer is not very easy. We would have to scan through the entire file to try to find anything. If we are going to use an accounting package or reporting software, CSV may be exactly what we need. However, if we wish to use it in a more ad hoc sort of way, it is not very readable.

Database CDR logging

If we wish to read our CDR logs, it is most easily accomplished when the records are sortable. The easiest way to do this is to store our CDR records in a database.

Even in this, Asterisk gives us choices. Included with Asterisk is support for PostgreSQL databases. In order to be able to install this, we must first have the `postgresql-devel` package installed on our system. If you have to install this package, you'll need to reinstall Asterisk. The automake system will automatically detect that we have the capability to use PostgreSQL and compile that module for us.

Quality Assurance

Aside from the development packages we have installed, we will also need a PostgreSQL server somewhere in our network. It can be the same machine as the Asterisk server, but it doesn't necessarily need to be. In fact, it probably makes sense to have only one such database server on our network, and we don't want to tie up too much of our PBX's resources with database maintenance and storage.

There is a script in `/usr/src/asterisk/contrib/scripts/` called `postgres_cdr.sql`, which creates the correct table structure for us. This script should be run from the database server.

If we get an error message while rebuilding that says something like "cannot find -lz", then we need to install `zlib-devel`.

Now that we have set up our database and installed the CDR module, we must configure Asterisk to use the correct database. In order to do this, we need to edit `/etc/asterisk/cdr_pgsql.conf`. All of the configuration variables are in the global section. Our file should look like the following:

```
[global]
hostname=dbserver.mydomain.tld
port=5432
dbname=asterisk
password=supersecret
user=asteriskuser
```

Once we have these variables set, the next time we restart Asterisk, all CDR records will be logged in the database.

If PostgreSQL is not our database of choice, we can use MySQL. This is not a part of the normal distribution of Asterisk. But as we have already installed `asterisk-addons`, we should already have the ability to use MySQL for CDR logging.

Before we compile, we need to make sure that we have `mysql-devel` installed. First, we need to decide which version we're going to use. Because of some license quibbles, MySQL version 4.0 and later is not in the automatic package distribution chain. Instead, if we do need to download it, we will have to get it directly from www.mysql.com. However, the older version (3.x) will work with Asterisk and hence, you may wish to take a look at the differences between what version 3 offered and what later versions give us.

Other than the development package mentioned, we will also need a MySQL server somewhere in our network. Just as with PostgreSQL, we can choose to have it on the same server as Asterisk, but for the same reasons, we probably shouldn't.

Next, on the database server, we need to create the database with a user and a table for the CDR data. We do this by running the following code:

```
# mysqladmin create database asteriskcdrdb

# mysql

mysql> GRANT ALL PRIVILEGES
    -> ON asteriskcdrdb.*
    -> TO asteriskcdruser
    -> IDENTIFIED BY 'changethis2yourpassword';

mysql> USE asteriskcdrdb;
mysql> CREATE TABLE cdr (
    -> uniqueid varchar(32) NOT NULL default '',
    -> userfield varchar(255) NOT NULL default '',
    -> accountcode varchar(20) NOT NULL default '',
    -> src varchar(80) NOT NULL default '',
    -> dst varchar(80) NOT NULL default '',
    -> dcontext varchar(80) NOT NULL default '',
    -> clid varchar(80) NOT NULL default '',
    -> channel varchar(80) NOT NULL default '',
    -> dstchannel varchar(80) NOT NULL default '',
    -> lastapp varchar(80) NOT NULL default '',
    -> lastdata varchar(80) NOT NULL default '',
    -> calldate datetime NOT NULL default '0000-00-00 00:00:00',
    -> duration int(11) NOT NULL default '0',
    -> billsec int(11) NOT NULL default '0',
    -> disposition varchar(45) NOT NULL default '',
    -> amaflags int(11) NOT NULL default '0'
    -> );
```

That's all there is to it! We only have to do this once, so it's really not so bad. Next, we have to modify the `/etc/asterisk/cdr_mysql.conf` file to correctly reflect our choices.

```
[global]
hostname=ourdbserver.ourdomain.tld
dbname=asteriskcdrdb
password=changethis2yourpassword
user=asteriskcdruser
port=3306
userfield=1
```

The next time we restart Asterisk, our CDR information will be stored in the MySQL database. What does that give us? We now have the ability to use a number of very powerful tools to search our CDR records to find trends and patterns.

Monitoring calls

Slightly less friendly than recording the information about a call is enabling the ability to monitor calls in real time. This allows us to listen in to a conversation as it happens, so that we may see how our customers are being treated.

The application to use to monitor a DAHDI channel is called `DAHDIBarge`. It can only accept one command-line argument, which is the number of the channel to listen in on. If we do not pass DAHDIBarge an argument, it will prompt us to enter one. The channel numbers it requests are the same channel numbers given in `system.conf` and `chan_dahdi.conf`.

Suppose we had four outgoing DAHDI channels, numbered 1 through 4. We could have something like this in our `extensions.conf` file:

```
exten => 8700,1,DAHDIBarge
exten => 8700,2,Hangup
exten => 8701,1,DAHDIBarge(1)
exten => 8701,2,Hangup
exten => 8702,1,DAHDIBarge(2)
exten => 8702,2,Hangup
exten => 8703,1,DAHDIBarge(3)
exten => 8703,2,Hangup
exten => 8704,1,DAHDIBarge(4)
exten => 8704,2,Hangup
```

This way, extension `8700` would give us access to any DAHDI channel; whether it is an FXO or FXS interface does not matter. Then, extensions `8701` through `8704` would give access to each of the outgoing interfaces.

Monitoring calls used to only be available via DAHDI (earlier known as Zaptel). However, in recent releases of Asterisk, call monitoring is now available for SIP channels. The application to use to monitor a SIP channel is called **ChanSpy**.

In order to use ChanSpy, the following format can be applied in your `extensions.conf` file: `ChanSpy([<chanprefix>][,<options>])`.

For example:

```
exten => 556,1,ChanSpy(scan)
```

The following are list of options available when executing ChanSpy:

- `-b`: Only spy on channels involved in a bridged call.
- `-g(grp)`: Match only channels where their `${SPYGROUP}` variable is set to contain `grp` in an optional, delimited list.

- `-q`: Don't play a beep when beginning to spy on a channel, or speak the selected channel name.
- `-r[(basename)]`: Record the session to the `monitor spool` directory. An optional base for the filename may be specified. The default is `chanspy`.
- `-v([value])`: Adjust the initial volume in the range from -4 to 4. A negative value refers to a quieter setting.

Since 1.4:

- `-w`: Enable 'whisper' mode, so the spying channel can talk to the spied-on channel.
- `-W`: Enable 'private whisper' mode, so the spying channel can talk to the spied-on channel but cannot listen to that channel.

Since 1.6:

- `-o`: Only listen to audio coming from this channel.
- `-X`: Allow the user to exit ChanSpy to a valid single digit numeric extension in the current context or the context specified by the `SPY_EXIT_CONTEXT` channel variable. The name of the last channel that was spied on will be stored in the `SPY_CHANNEL` variable.
- `-e(ext)`: Enable 'enforced' mode, so the spying channel can only monitor extensions whose name is in the `ext` delimited list.

The following actions may be performed when using ChanSpy:

- Dialing # cycles the volume level.
- Dialing * will stop spying and look for another channel to spy on.
- Dialing a series of digits followed by # builds a channel name to append to `<chanprefix>`. For example, run `ChanSpy(Agent)` and dial `1234#` while spying to jump to channel `Agent/1234`.

While these extensions are useful, there is a danger, and we must consider security. Clearly, we do not want just any employee to be able to listen to calls that are in progress. And more than that, we really don't want our customers to be able to accidentally listen to calls that are in progress.

A good way to handle this problem is to create a separate context just for monitoring extensions. Then, designate a single telephone handset that will be able to do nothing but monitor extensions. This handset should be the only phone in the monitoring context, and the monitoring context should not be included in any other context. Keep that handset under lock and key. Not only will this keep people from overhearing embarrassing or confidential information, it will also go a long way towards fostering trust with the employees.

Recording calls

The last of all of the quality assurance methods we will discuss is the call recording capability of Asterisk. This is highest on the Big Brother chart because phone conversations can be archived forever and reviewed on demand. Therefore, an employee's entire telephone history can be called up at any time.

This feature can be accessed from a number of different sources. First, we can configure specific call queues to record calls. This is done in the `queues.conf` file, for each individual queue. We set it thus:

```
[100]
. . .
monitor-format = wav
monitor-join = yes
```

The first line tells Asterisk to record the conversation in the `.wav` format. This is the best choice because it is most compatible with other operating systems. As archived conversations can be burned to CDs, compatibility is a high priority. The second line tells Asterisk to join the two files (in and out) into one file. If we do not do this, we will only hear half of the conversation. In order to take advantage of this feature, we must have `soxmix` installed on our Asterisk server. The Red Hat packages that contain `sox` are missing `soxmix`. Therefore, in order to install `soxmix` on Red Hat Linux, we need to do so from source.

All calls coming into the queue will be recorded. The name of the file will be the unique ID that Asterisk generates for every call. If we wish to change this, we can do so by adding something like the following in `extensions.conf`:

```
exten => 100,1,SetVar(MONITOR_FILENAME=${DATETIME}-${CALLERID(num)})
exten => 100,2,Queue(100)
```

This will record all calls coming through the queue named `100` in `.wav` format. We are then free to encode them into MP3 format if we wish to save space.

Aside from recording calls in queues, we can also monitor arbitrary calls through the use of the dialplan. The name of the application that records a channel is `Record`. In order to start recording, we call the application as follows:

```
exten => 200,1,Record(${TIMESTAMP}${CALLERID(num)}-${EXTEN}.wav)
exten => 200,2,Dial(SIP/1001)
```

With just one line of code in our dialplan, we can start monitoring calls. If we want, we could even insert this line into our macro definitions for standard extension types. Or, we could do something like the following:

```
[incoming]
exten => _.,1,Record(${TIMESTAMP}${CALLERID(num)}-${EXTEN}.wav)
exten => _.,2,Playback(thiscallmaybemonitoredorrecorded)
exten => _.,3,Goto(default,${EXTEN},1)
```

In three lines of code, we have enabled recording for all incoming phone calls. We have even notified our customers that the call may be recorded. We have the power. Should we use it?

Legal concerns

> This is not legal advice. Only a qualified attorney can advise you on your particular situation.

It is very important to note that, just because we can monitor calls, doesn't mean we should, or even that it would be legal to do so. Many states in the United States of America are two-party or all-party states. This means that all parties to a conversation must know that a call is being recorded for it to be admissible in court.

More than that, there are privacy laws in place to protect everyone. Only a careful study of all applicable laws can tell us if we are in the clear. We should never record any phone calls until we have spoken to a lawyer.

For those of you hosting a PBX server, you might just want to play it safe and restrict call recording or monitoring. However, there are alternative solutions for those who wish to individually record conversations indepent of the Asterisk server. Counterpath softphone products such as X-Lite has a call recording option, which will save the recordings to your local computer and in WAV format. There is also a hard phone solution through the Polycom 650/670 IP phone where you can plug in a USB memory stick into the phone. Once plugged in you can enable call recording and your conversation will be saved directly to the memory stick.

But aside from the legal issues, there are also moral issues. Maybe it depends on our intent when we call. Are we recording the calls to help our employees improve? Are we recording the calls so that we have an accurate representation of what was agreed upon? Or are we recording calls to try to trap someone, or to pull information out of calls to be used out of context later?

Summary

As we have seen in this chapter, Asterisk gives us the power to:

- Record call information (CDR)
- Monitor conversations (ChanSpy)
- Record the conversations themselves

The purpose of these capabilities is to provide us with options for using our system effectively. It is our responsibility to use these powers appropriately.

There is no point recording all calls if you are never going to use those recordings. Similarly, a database is an overkill if you have no real interest in your calling history.

However, there are many reasons to use these features. For instance, to produce reports or answer questions that other users or departments have regarding the telephone system. The users of the system will know more about what they will need in order to carry out their day-to-day duties, which is why we spend time figuring out exactly what they need early in the deployment plan—to ensure the system provides everything that is needed.

7
Making Asterisk Easy to Manage

Now that we have covered the basics, you have probably noticed that Asterisk by itself may not be so easy to manage due to the nature of its flat file (CONF files) structure. For example, if you delete an extension, you are likely to find yourself opening multiple CONF files to remove references and settings pertaining to that extension.

The good news is that there are free open source applications that have been developed. These implement a GUI for Asterisk, as well as a relational database (MySQL), which stores your settings and updates the CONF files automatically.

One such application is called FreePBX and can be downloaded from `www.freepbx.org`. FreePBX also packages several other applications such as a voicemail portal for accessing voicemail and extension settings, and an operator panel to monitor and oversee who is on the phone. It also provides other useful utilities such as a call detail report interface to retrieve call records as well as identify calling patterns and traffic growth. Later in this chapter, we will go over the different sections and applications belonging to FreePBX.

There are also all-in-one open source solutions that build and set up your operating system, Asterisk service, database server, web server, and much more, all within an hour or less. One of the most popular of these solutions is Trixbox CE.

Trixbox

As we discussed when introducing Asterisk, flexibility is a primary focus. Asterisk can be used for a variety of different purposes and each available feature can be tailored to the specific needs of any organization. Trixbox retains some of the flexibility and adds a massive amount of convenience and ease of use. It offers web-based configuration, web access to voicemail, reporting, and other functions, which we will cover shortly. All these functions can be added to your own Asterisk installation as they are all based on existing Asterisk tools. The real benefit of Trixbox is that you don't have to set these up manually to make them work together. Trixbox installs and sets up the base configuration for all of the tools it provides.

It's extremely easy to set up and use. The downside to this is that you lose some of the flexibility. For example, you can't choose which OS to base your Asterisk system on and you can't fully customize the configuration, which becomes a hindrance in larger Asterisk installations.

CentOS

Trixbox is designed around the CentOS distribution of Linux. CentOS is built from the Red Hat Enterprise source packages. It has a relatively small core team of developers that concentrate on packaging the OS without Red Hat's proprietary components. The main focus of CentOS is to provide a freely available operating system with the packages and features needed at enterprise level, without the cost associated with the base distribution—Red Hat Enterprise Linux. However, CentOS does offer a range of commercial support, which is invaluable to most enterprises and thus is an option we can consider.

CentOS isn't the focus of this chapter and it doesn't really have too much bearing on our use of Trixbox other than knowing basically how to use and update it. We will focus on the setup and maintenance of Trixbox and the features it provides for us. If we decide to use Trixbox, it would be beneficial to spend time getting to know CentOS.

Trixbox preparation and installation

Trixbox recommends a minimum of a 2.4 GHz processor and 512 MB of RAM. However, given the decline in price of hardware components as well as your possible need to enable certain resource-demanding features such as call recording, you might want to get 2 GB of RAM and if possible, a Dual-Core CPU. The amount of RAM required has a direct correspondence with how heavily used the system will be. As Trixbox is Asterisk with a few other services added, we can pretty much scale it similarly. We do have to consider extra resources for the additional services we have running, such as the web server and the MySQL server.

Trixbox comes in an ISO image that can be burned and installed as a full OS. The ISO installs a modified CentOS system automatically and sets up the necessary Trixbox services.

We can obtain the ISO from `http://www.trixbox.org/downloads`.

After downloading and burning the image to disk, reboot the target machine with the Trixbox CD in the drive, and wait for the prompt, which should look like the following:

```
boot: _
```

If at this point we hit *Enter*, the installer will start and begin to install CentOS with Trixbox on the first primary hard disk. It's very important to ensure that this disk is the disk we want to use and that no important data is held there, as all data will be lost. From this point onwards, installation is entirely automatic. We can now leave it for a few minutes while it prepares the machine, installs the OS and the necessary programs for Trixbox—including Asterisk, MySQL, Apache, and so on. It's a good time to gather the documents we need to configure Asterisk, such as our lists of extensions and our service provider account details. We'll need the same information as in the previous chapters, where we set up Asterisk manually. However, now we don't need to worry so much about Asterisk's configuration syntax as we have a friendly GUI-based setup system that takes care of most things.

Installation of Trixbox is extremely simple and as long as all of our hardware has Linux support, there should be little issue getting the system installed.

We can configure advanced options and modify the kernel boot parameters if necessary by hitting any one of the keys from *F1* to *F5* at the boot prompt (this usually isn't necessary). *F5* is of particular note as this runs the CD as a rescue disk, which we can use to repair a machine that refuses to boot.

Also, if we have problems getting the OS installed from the CD, we can enter `linux mediacheck` at the boot prompt to confirm the integrity of the installation disk. This is something worth doing to ensure that the ISO was burnt properly rather than waiting for it to break during installation.

When the system has finished installing, it should reboot and leave us at the login prompt. We should log in as root and change the default system passwords. The commands we need to run are:

```
# passwd            ;to change our root password
# passwd-maint      ;to change the maint user's password for the Asterisk
                    Management Portal
# passwd-meetme     ;to change the Web MeetMe user's password
```

It's important we change these passwords as soon as possible, to ensure that we don't deploy the system with default passwords, in order to avoid the risk of a possible security breach.

After we have the passwords set, it's time to ensure that we can access the machine over the network. Firstly, we should check if the machine has picked up an IP address at boot. We should check if we have an IP at login, but to be absolutely sure, we should run `ifconfig` and check that an IP address has been assigned to our required network interface.

```
# ifconfig eth0 | grep "inet addr"
```

This should show a line containing our IP address. If no IP address is shown or we want to set a static IP for the Asterisk box, which is often more useful, then we can run `system-config-network`. In this case, we will have to input the network settings for the machine, namely the IP address we want to use, the subnet mask, default gateway, and DNS server.

```
# system-config-network             ;Enter IP details
# /etc/init.d/network restart  ;Apply the changed settings
```

> **IP addressing**
>
> As we may have SIP clients and as we access Trixbox CE using a web browser, it is usually beneficial to have the Trixbox CE machine configured with a static IP, or if we are using DHCP, to ensure the address is reserved so that it doesn't change. This ensures we can always find the server without reconfiguring clients.

What is FreePBX?

Trixbox is an excellent solution for those who want to get an Asterisk solution up and running fast. However, for those who wish to choose their own OS or have the flexibility of deciding what applications and services to install, you might want to go with FreePBX. FreePBX is an easy-to-use GUI (Graphical User Interface) that controls and manages Asterisk. For a while, Trixbox used to bundle FreePBX as the primary configuration interface for Trixbox. Even today much of the source code that Trixbox uses for its PBX settings is based on FreePBX. Therefore, as we go forward in exploring the FreePBX GUI, Trixbox users can easily follow as the interfaces belonging to these two solutions are almost identical.

FreePBX preparation and installation

As mentioned, FreePBX gives you the flexibility of choosing your OS. Given that CentOS is one of the most popular OS distributions for Asterisk, we will demonstrate the preparation and installation of FreePBX based on CentOS 5.1. To make things easier, the instructions below will also recap installing Asterisk.

1. Install CentOS, enabling the following packages:
    - DNS Server
    - Web Server
    - Mail Server
    - MySQL Database
    - Development Tools

    Use the following code to enable these packages:

    ```
    yum install nano
    reboot
    ```

2. **Edit network settings**: Most installations have their network settings set to DHCP by default. As we do not want the internal IP address of our server to change, its best to statically assign an IP address.

    ```
    nano /etc/sysconfig/network
    HOSTNAME=internal.hostname.DOMAIN.com (Set your internal hostname name here)
    ```

    Press *Ctrl+X* to save, *Y* to confirm.

    ```
    nano /etc/sysconfig/network-scripts/ifcfg-eth0
    IPADDR=192.168.1.20
    NETMASK=255.255.255.0
    GATEWAY=192.168.1.1
    NETWORK=192.168.1.0
    ```

    Note that the above information is just an example. You should replace the address above with address settings that match your network.

    Press *Ctrl+X* to save, *Y* to confirm.

    ```
    echo "options {" >> /etc/named.conf
    echo "    directory \"/var/named\";" >> /etc/named.conf
    echo "    dump-file \"/var/named/data/cache_dump.db\";" >> /etc/named.conf
    echo "    statistics-file \"/var/named/data/named_stats.txt\";" >> /etc/named.conf
    ```

```
echo "};" >> /etc/named.conf
echo "include \"/etc/rndc.key\";" >> /etc/named.conf

service named start

chkconfig named on

nano /etc/resolv.conf

search internal.DOMAIN.com (Set your internal domain name here)
nameserver 192.168.1.5
nameserver 127.0.0.1

nano /etc/hosts

127.0.0.1 internal.hostname.DOMAIN.com (Set your internal hostname
name here)
127.0.0.1 asterisk1.local
127.0.0.1 localhost
```

Press *Ctrl+X* to save, *Y* to confirm.

```
iptables -P INPUT ACCEPT
iptables -P OUTPUT ACCEPT
iptables -P FORWARD ACCEPT
iptables -F
iptables -X

/etc/init.d/iptables save

service network restart
```

3. **To Update**:

   ```
   yum -y update
   ```

4. **To Disable Selinux**:

   ```
   echo "selinux=disabled" > /etc/selinux/config
   reboot
   ```

5. **Install dependencies and extra packages**:

   ```
   yum install e2fsprogs-devel keyutils-libs-devel krb5-devel libogg
   libselinux-devel libsepol-devel libxml2-devel libtiff-devel
   gmp php-pear php-pear-DB php-gd php-mysql php-pdo kernel-devel
   ncurses-devel audiofile-devel libogg-devel openssl-devel mysql-
   devel zlib-devel perl-DateManip sendmail-cf sox

   cd /usr/src
   wget http://easynews.dl.sourceforge.net/sourceforge/lame/lame-
   3.97.tar.gz
   tar zxvf lame-3.97.tar.gz
   ```

```
cd lame-3.97
./configure
make
make install
```

6. **Install Asterisk and FreePBX**:

   ```
   cd /usr/src

   wget http://downloads.asterisk.org/pub/telephony/asterisk/
   asterisk-1.6.1-current.tar.gz wget http://downloads.asterisk.org/
   pub/telephony/asterisk/asterisk-addons-1.6.1-current.tar.gz wget
   http://downloads.digium.com/pub/telephony/dahdi-linux-
   complete-current.tar.gz
   wget http://downloads.digium.com/pub/libpri/libpri-1.4-current.
   tar.gz
   wget http://mirror.freepbx.org/freepbx-2.5.0.tar.gz

   tar zxvf asterisk-1.6.1-current.tar.gz
   tar zxvf asterisk-addons-1.6.1-current.tar.gz
   tar zxvf dahdi-linux-complete-current.tar.gz
   tar zxvf libpri-1.4-current.tar.gz
   tar zxvf freepbx-2.5.0.tar.gz

   cd /var/lib/asterisk/sounds
   wget http://downloads.digium.com/pub/telephony/sounds/asterisk-
   extra-sounds-en-gsm-current.tar.gz
   tar zxvf asterisk-extra-sounds-en-gsm-current.tar.gz

   cd /usr/src/dahdi-linux-complete-CURRENT

   make
   make install
   make config
   /sbin/ztcfg

   echo "/sbin/ztcfg" >> /etc/rc.d/rc.local

   cd /usr/src/libpri-1.4-CURRENT

   ./configure
   make menuselect
   make
   make install

   cd /usr/src/asterisk-1.6.1-CURRENT

   useradd -c "Asterisk PBX" -d /var/lib/asterisk asterisk
   mkdir /var/run/asterisk
   mkdir /var/log/asterisk
   ```

```
chown -R asterisk:asterisk /var/run/asterisk
chown -R asterisk:asterisk /var/log/asterisk
chown -R asterisk:asterisk /var/lib/php/session/

nano +231 /etc/httpd/conf/httpd.conf
```

Change User apache and Group apache to User asterisk and Group asterisk.

Ctrl+X to save, *Y* to confirm.

```
nano +329 /etc/httpd/conf/httpd.conf
```

Change `AllowOverride` None to `AllowOverride` All.

Press *Ctrl+X* to save, *Y* to confirm.

```
./configure
make
make install

/etc/init.d/mysqld start

cd /usr/src/freepbx-2.5.0

mysqladmin create asterisk
mysqladmin create asteriskcdrdb
mysql asterisk < SQL/newinstall.sql
mysql asteriskcdrdb < SQL/cdr_mysql_table.sql

mysql

GRANT ALL PRIVILEGES ON asteriskcdrdb.* TO asteriskuser@localhost
IDENTIFIED BY 'SOMEPASSWORD';
GRANT ALL PRIVILEGES ON asterisk.* TO asteriskuser@localhost
IDENTIFIED BY 'SOMEPASSWORD';
flush privileges;

\q

mysqladmin -u root password 'SOMEPASSWORD'

cd /usr/src/asterisk-addons

./configure
make
make install

cd /usr/src/freepbx-2.5.0

./start_asterisk start

yum install php-pear-DB
yum install php-mysql
```

```
./install_amp --username=asteriskuser --password=SOMEPASSWORD
echo "/usr/local/sbin/amportal start" >> /etc/rc.local
chkconfig httpd on
chkconfig mysqld on
```

Open your browser and type `http://ipaddressofpbx/admin`.

Click on the red bar in FreePBX.

7. **Fix ARI password**:

   ```
   nano -w /var/www/html/recordings/includes/main.conf.php
   $ari_admin_password = "SOMEPASSWORD";
   ```

   Press *Ctrl+X* to save, *Y* to confirm.

8. **Configure Sendmail**:

   ```
   nano /etc/mail/sendmail.mc
   define(`SMART_HOST', `relay.DOMAIN.com)dnl
   MASQUERADE_AS(`pbx.DOMAIN.com')dnl
   FEATURE(`masquerade_envelope')dnl
   ```

   Press *Ctrl+X* to save, *Y* to confirm.

   ```
   make -C /etc/mail
   ```

9. **Edit** `sip_nat.conf` **for proper NAT**:

   ```
   nano /etc/asterisk/sip_nat.conf

   localnet=192.168.1.0/255.255.255.0
   externhost=pbx.DOMAIN.com (Set your external hostname name here)
   externrefresh=10
   fromdomain=DOMAIN.com (Set your external domain name here)
   nat=yes
   qualify=yes
   canreinvite=no
   ```

   Press *Ctrl+X* to save, *Y* to confirm.

10. **Add extra codecs to config**:

    ```
    nano /etc/asterisk/sip_custom.conf
    allow=gsm
    ```

Making Asterisk Easy to Manage

```
allow=h261
allow=h263
allow=h263p
videosupport=yes
```

Press *Ctrl+X* to save, *Y* to confirm.

```
nano /etc/asterisk/iax_custom.conf

allow=gsm
allow=h261
allow=h263
allow=h263p
videosupport=yes
```

Press *Ctrl+X* to save, *Y* to confirm.

```
asterisk -rx reload
```

11. **Edit voicemail config**:

    ```
    nano /etc/amportal.conf
    ```

    If the web interface on your PBX will be accessible on the internet:

    ```
    AMPWEBADDRESS=pbx.DOMAIN.com (Set your external hostname name here)
    ```

    If the web interface on your PBX will be accessible only on your internal network:

    ```
    AMPWEBADDRESS=internal.hostname.DOMAIN.com (Set your internal hostname name here)
    ```

    Press *Ctrl+X* to save, *Y* to confirm.

    Or if your users will NOT have access to the web interface:

    ```
    nano /etc/asterisk/vm_email.inc
    ```

    Remove `Visit http://AMPWEBADDRESS/cgi-bin/vmail.cgi?action=login&mailbox=${VM_MAILBOX} to check your voicemail with a web browser.\n.`

    Press *Ctrl+X* to save, *Y* to confirm.

```
nano /etc/asterisk/vm_general.inc
serveremail=pbx@DOMAIN.com ; Who the e-mail notification should
appear to come from
fromstring=DOMAIN PBX ; Real name of email sender
```

Press *Ctrl+X* to save, *Y* to confirm.

12. **Fix MOH directory**:

    ```
    ln -s /var/lib/asterisk/moh /var/lib/asterisk/mohmp3
    asterisk -rx reload
    ```

13. **Open your browser and type** `http://ipaddressofpbx`.

It's done!

FreePBX System Status Dashboard

Now that our system is installed, base passwords are set, and we have network connectivity, we can begin to configure the server to perform its role. This is done by using the web management utilities the system provides, and in some cases when necessary by modifying the underlying Asterisk configuration files.

We will take a look at the functionality provided with the web interface and then follow an example setup, which will create a working Asterisk server with a single PSTN line and a single SIP extension.

The first place we want to look is most likely the **FreePBX System Status Dashboard**, which allows us to configure most of the features of the Asterisk system in an eye-pleasing and user-friendly fashion. With this interface, we can manage everything including extensions, ring groups and trunks, our MySQL database, and even report generation by the system.

Making Asterisk Easy to Manage

To get to the FreePBX System Status Dashboard, open a web browser and type `http://<AsteriskIP>/admin` in the address bar (use the IP address we set earlier). Once this is entered, you will be prompted to enter a username and password. By default, the username is "maint" and the password is "password". After you log in, you should be presented with a screen like this:

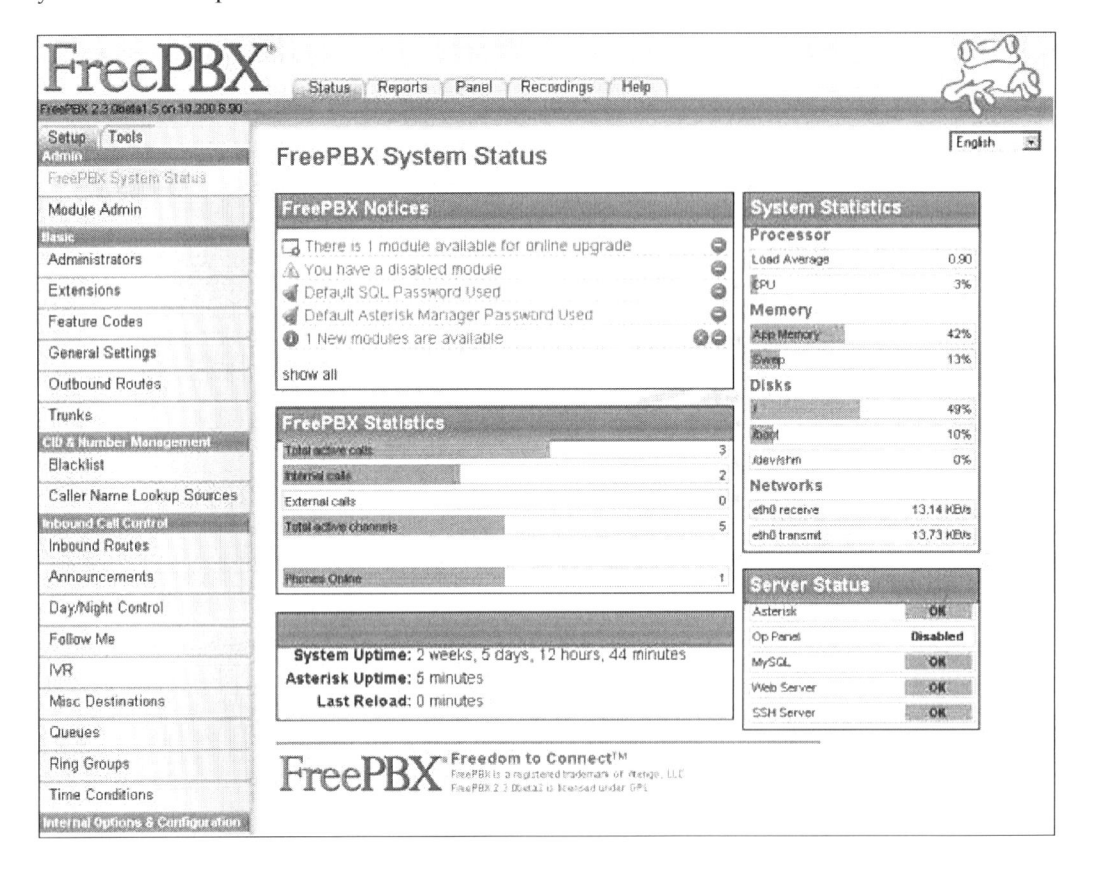

Tools

This section is where we can maintain the system, check the status of services, perform maintenance tasks such as backup and restore, as well as access third-party add-ons.

- **System Status**: This page shows the current status of the system. It should show whether or not Asterisk, cron, secure shell, and the web server are running. It will also give us a brief overview of how many channels are active as well as how many extensions and trunks are connected. There is even a **System Statistics** section showing an overview of the systems resources.

- **Print Extensions**: This section creates a printable list of names and extension numbers.

- **Asterisk CLI**: This interface will give you the ability to send Asterisk commands directly through the GUI.

- **Asterisk API**: If you have custom applications or are using third-party clients, which need to communicate with Asterisk, this section will allow you to create and assign privileges to Asterisk.

- **Backup & Restore**: You can configure a regular backup schedule to ensure that you have a copy of your Asterisk and FreePBX configuration, voicemail, and CDR records. You can also restore a previous backup, in case of data loss or a major configuration fault. Backups are stored on the file system at `/var/lib/asterisk/backups`. You should make it a point to maintain an offline copy of important backups.

- **Asterisk Log Files**: This section will allow you to see the last 2000 lines of the Asterisk debug log. This can prove very useful when trying to troubleshoot issues with the Asterisk service.

- **Asterisk Info**: This is overview information that focuses on Asterisk. We can see information about SIP, Zaptel, and IAX usage as well as version, and system information for the Asterisk service. This section can prove invaluable in troubleshooting, as this section also specifies the IP address of a connected extension as well as its latency.

Setup

Under this section, we will find the relevant pages for configuring our extensions, lines, trunks, conferences, and other Asterisk features.

- **Inbound Routes**: Here, we can configure how incoming calls from the PSTN are handled. We can route them to a receptionist, extension, ring group, IVR, or queue.

- **Extensions**: We would use this section to add an extension. The information we would provide would be—the protocol in use (for example, SIP), the extension number, password, the user's full name, and whether or not we would like to record incoming and/or outgoing calls for this extension. In this section you can also specify other parameters such as outbound caller ID, which defines what phone number will appear to the person you are calling.

    This obviously is simpler and more intuitive than modifying the Asterisk configuration files directly. Although we do lose a little bit of flexibility in how these extensions are added, we can make up for this with some hand-hacking as required. We can also configure voicemail for the extension here if we wish.

- **Ring Groups**: If we need a group of extensions to ring together or in a given pattern, as covered in a previous chapter, we need to set up ring groups. This section lets us create groups and detail the extensions that we would like to have in these groups. Ring groups are often used when you have a large number of agents and a moderate call volume. Therefore, a ring group is used when there is an expectation that the caller will be answered within 60 seconds or less. If for whatever reason all agents are busy or the maximum wait time is exceeded, this section will also allow you to specify a failover destination in which you can choose to fail over the call to another ring group, queue, extension, voicemail box, IVR, or any other FreePBX module.

- **Queues**: If we expect large volumes of calls, we will need to queue them and this section is where we configure our queues. In order to configure a queue, we provide the ID, name, password, CID prefix, and available agents. We can also add on-hold music and set other queue options such as announcements and their frequency. Similar to ring groups, here also we can define a failover destination if need be.

- **IVR**: An IVR can be used for failover when a queue is at capacity or an extension isn't available. Some companies simply like to offer callers a greeting in which they can choose the department or extension they desire without the need of a receptionist. The caller can even be routed to a company directory.

- **Trunks**: This allows us to add a variety of trunk types such as SIP and IAX to our system, much as we did in previous chapters. A nice wizard-based system will prompt us for all the necessary parameters such as trunk type, name, DIDs, and so on.

- **Outbound Routes**: We should now be aware whether calls to different places should be routed over different providers. For example, we may have a local service and a VoIP server where we use VoIP for international calls and the local service for local and national calls. In this section, we can configure dial patterns to ensure we route calls over the most efficient and cost-effective line.

- **On hold Music/System Recordings**: We can record, upload, and manage our various audio files for use with the system. You can create multiple hold music categories, which comes in handy when creating queues and having different hold music for different queues. The system will accept either WAV or MP3. However, it is recommended to use WAV files.

- **General Settings**: Here we can configure settings such as the number to dial for an outside line and a fax machine extension or email address. In this section, we can also define whether we can use the company directory option to search by first name, or last name, or both. In this section you can also specify Asterisk dial command options such as allowing callers who don't have a transfer button to initiate a transfer through the # key.

- **Time Conditions**: In this section, we can specify different time periods and then route callers to different destinations depending on either the time, day, week, or month. For example, a company can play one IVR during 9 A.M. to 5 P.M. and then a different IVR after these hours.

Trixbox maintenance section

For those using Trixbox, the maintenance page can be accessed through `http://<AsteriskIP>/maint` and to directly get to the PBX settings section (almost identical to the FreePBX interface) enter `http://<AsteriskIP>/admin`.

This section offers a great deal of useful applications developed by the Trixbox CE team. These applications are not normally found in FreePBX.

- **Endpoint Manager**: This section enables administrators to easily discover VoIP phones on the network and configure them with an extension. Trixbox also acts as a TFTP server, so all you need is the make, model, and MAC address of the VoIP phone. Then just specify the extension you want to assign to that phone. Once all this is completed, simply point your VoIP hardware towards the IP address of the Trixbox server and your device will download all the settings necessary to register an extension.

- **Config Edit**: This section lets us access the configuration files for Asterisk and CentOS through a web-based editor. For example, we can edit `Asterisk.conf`, `resolv.conf`, and any other files in the `/etc` directory. This becomes useful when we find an area of the GUI that doesn't fit our requirements for customizing the system and have to edit the relevant configuration file by hand.

- **phpMyAdmin**: This gives us access to the web-based MySQL management tool phpMyAdmin. This tool is extremely useful for TrixBox CE as much of the configuration and logs are held in a MySQL database. We can back up the databases, run SQL queries to view or modify the existing databases, and even add databases of our own.

- **Sysinfo**: This page gives an overview of the current system state—covering network, memory and hard disk utilization, as well as some system specifics.

- **Packages**: This section allows you to easily browse and install updates or patches for CentOS or Trixbox.

- **BackUp**: Trixbox has created its own backup and restore utility. The utility will allow you to schedule backups, define what you want to backup, and even specify remote backup parameters through FTP.

- **Bulk Extensions**: This tool is very handy especially when administrators are migrating an existing infrastructure to Trixbox. If you have a large number of extensions, adding them one by one can take a great deal of time. Therefore, to make things easier, this section will allow administrators to set up extensions in bulk by using spreadsheets containing the basic information needed to create an extension. As most administrators keep spreadsheets of their employees as well as extensions, this process can save them hours of work.

The following third-party applications are bundled in both FreePBX and Trixbox: Reports, ARI, and FOP.

Reports

This section is very useful and enables us to view reports by date, view full call logs, compare calls, as well as monitor monthly and daily traffic. Your custom report results can be downloaded to an Excel spreadsheet or PDF. This function is provided by the Areski Asterisk-stat tool (http://areski.net/asterisk-stat-v2/).

Asterisk Recording Interface

Asterisk Recording Interface (ARI) is a user portal for users to easily view and listen to their voicemail, call monitor logs and recordings, and allow them to change their own common PBX settings (call forward and voicemail passwords). This application is created from original work based on ARI from Littlejohn Consulting.

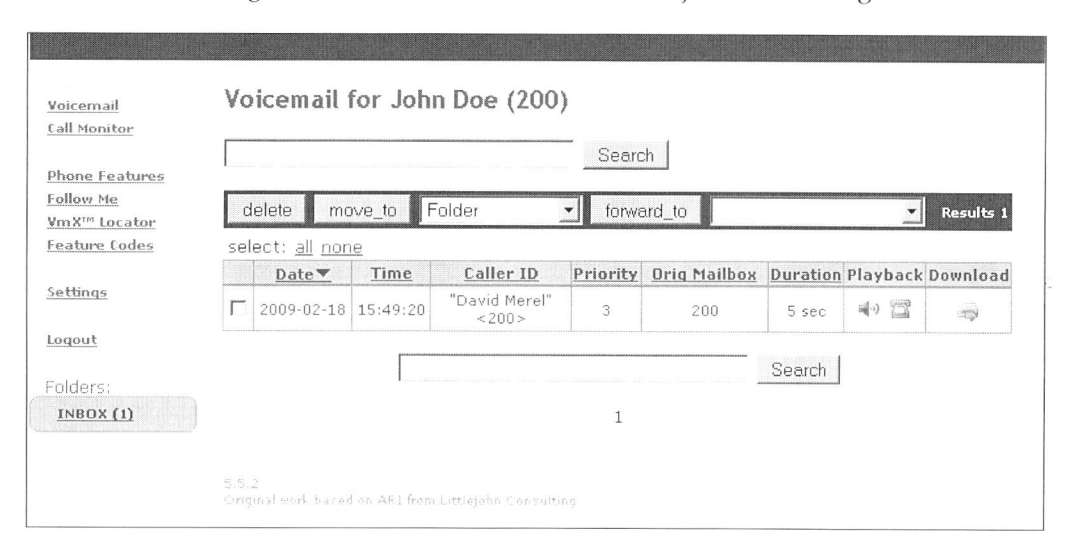

Flash Operator Panel (FOP)

This panel is extremely useful as it shows us all extensions, conferences, and queues with details of their status. We can use this to get a current overview of system usage. It's a Flash-based real-time interface to the system state. It can also be used to hang up, transfer, and originate calls through drag-and-drop, as well as provide pop-up functionality where the customer's details appear on screen according to their CLI details. All this can be protected so as to restrict agents' access to every function of the panel.

It's quite an intuitive interface, so most actions are taken with button clicks and mouse movements. For example, dragging a free channel to a bridged channel will allow us to barge into the existing call. Operators can also use this panel with drag-and-drop functionality in order to transfer calls from one extension to another. Imagine you want to transfer a call; through this panel you can simply drag the call from one extension to another extension without even lifting your phone's handset. This function is created by Asternic (http://www.asternic.org/).

A quick rundown of the Flash Operator Panel features is as follows:

At a glance you can see:

- Which extensions are busy, ringing, or available
- Who is talking and to whom (CLID, context, priority)
- SIP and IAX registration status and reachability
- MeetMe room status (number of participants)
- Queue status (number of users waiting)
- Message Waiting Indicator and count
- Parked channels
- Logged-in agents

You can perform these actions:

- Hang up a channel
- Transfer a call leg through drag-and-drop
- Originate calls through drag-and-drop
- Barge in on a call using drag-and-drop
- Set the caller ID when transferring or originating a call
- Automatically pop up a web page with customer details
- Click-to-dial from a web page
- Mute/Unmute MeetMe participants

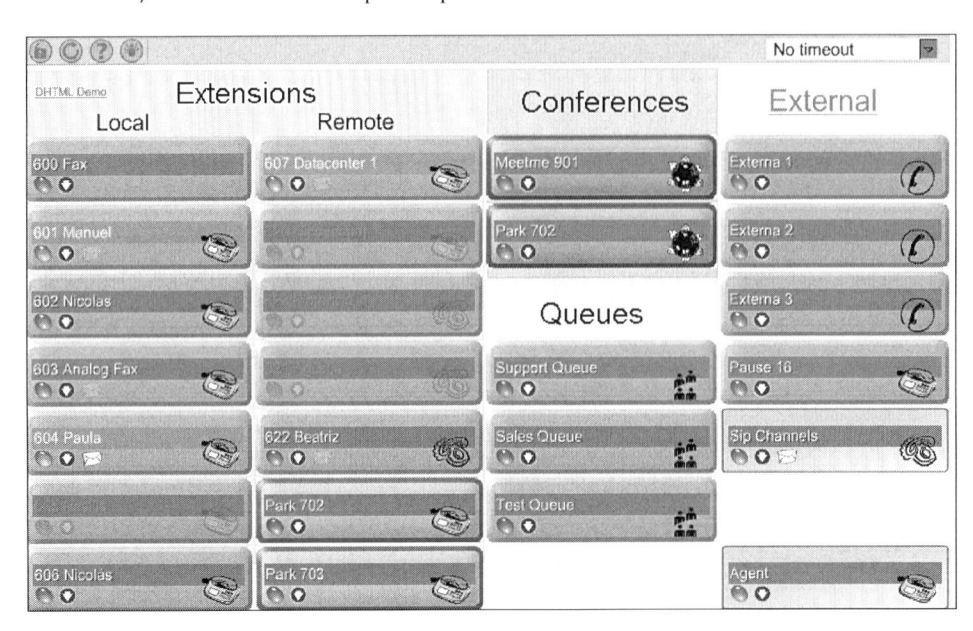

Flash operator configuration files

The **Flash Operator Panel (FOP)** can be configured by editing the configuration files that are shipped with it. They can be found in the `/var/www/html` folder and can also be accessed from the Asterisk Management Portal.

The files include:

- `op_astdb.cfg`
- `op_buttons.cfg`
- `op_buttons_additional.cfg`
- `op_buttons_custom.cfg`
- `op_server.cfg`
- `op_style.cfg`

`op_server.cfg` is the most important for initial setup. It contains the main FOP configuration, including the IP address of the Asterisk service, the username and password for accessing FOP, as well as any debug settings that should be applied. You can also configure your available conferences here. The other files can be used to add customized settings such as extra buttons for the system and modifications to the style of the FOP.

Web MeetMe

Web MeetMe is a web-based frontend to the MeetMe add-on for Asterisk. It allows us to monitor and control conferences.

MeetMe can be accessed through a web interface in which a user will enter their email address and password. Once authenticated, the user will be given the options to create a conference, delete or edit a conference, and monitor a conference (for those who are moderators). Moderators can listen into a conference, see an overview of participants, as well as mute or kick out participants. Creating a conference is pretty simple. All you need to do is specify a conference number (often one is randomly generated for you), a user PIN (for your participants), a moderator PIN (for the administrator or leader of the conference), and the time and date when the conference will be held.

There are several other options available such as the ability to invite participants, where the system will send out an email containing the conference information.

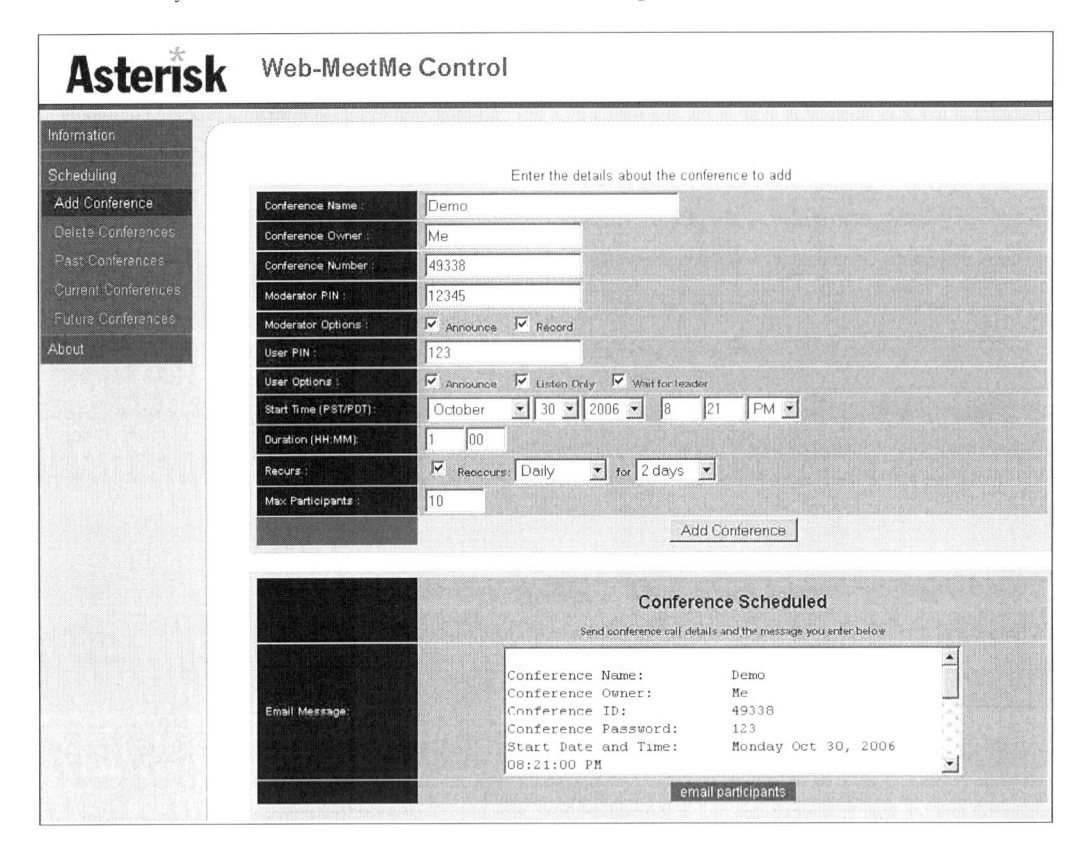

Currently WebMeetme 3.0 is pre-installed in **PBX in a Flash** which also uses FreePBX as the administration GUI.

For more information on PBX in a Flash, visit `http://pbxinaflash.net/`.

Alternatively, you can download and install MeetMe separately by going to: `http://sourceforge.net/projects/web-meetme/`.

Setting up and accessing Web MeetMe through Trixbox

Web Meet-Me 3 is included in Trixbox 2.4 (2.3.0.4 and above) but does require a few steps to be done before it will work properly.

In order to set up an extension to dial into for meetings please follow this example.

Under the **Tools** section, go to **Custom Destination** (you may have to download this module from the **Module admin** section also found under **Tools**).

In the **Custom Destination** field below, enter the following:

custom-meetme3,s,1

Enter **meetme** in the **Description** field. Then click on the **Submit Changes** button followed by the **Apply Configuration Changes** button.

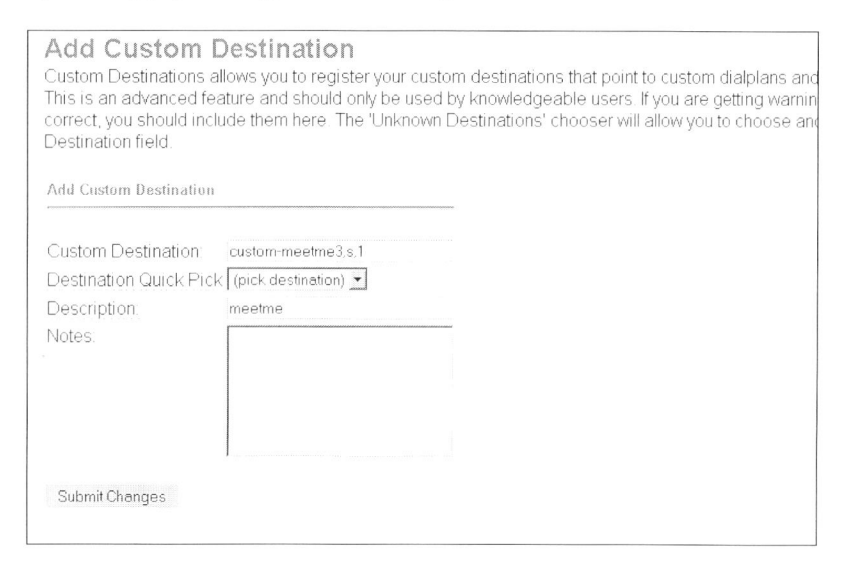

Now go to **Misc Applications** (you might have to install this module as it is not installed by default). Click on **Add Misc Application**. In the **Description** field, enter whatever you want.

meetme would be a good description. In the **Feature Code** field, put the extension you want users to dial to get into conferences. Then for **Destination**, select **Custom Destinations** and choose **meetme** from the drop-down list.

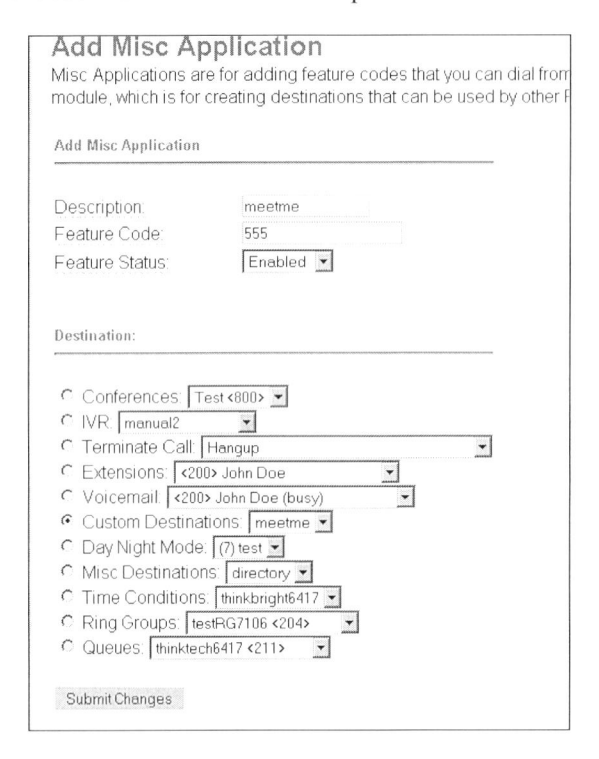

Then log in to MeetMe using the URL: `http://YourServer/web-meetme` and add a conference. There are two pre-configured users:

- Admin

    Username: wmm@localhost

    Password: wmmpw

- Standard user

    Username: tim@localhost

    Password: 1234

Log in with one of the accounts and click on **Add Conference**. Give it a name and conference number. This is the number you will need when you dial into the extension created in the **Misc Applications**. Click on the **Add Conference** button.

Now dial the extension you created in the Misc Applications step. You should be prompted for your conference number.

Flexibility when needed

We have looked at a few graphical configuration tools that add a lot of convenience and ease of use to the Asterisk system. As with any GUI, the focus is clarity, ease of use, and intuitive design. When we take a powerful command-line or service-based application and add a GUI to it, there is often a loss of flexibility. As Asterisk holds flexibility as one of its most important aspects, this may seem like a major downside to Trixbox/FreePBX.

However, we can still get under the skin and make up for some of the shortcomings in the graphical interfaces. As we have seen in the Trixbox Management Portal, you are provided with a direct link to the text configuration files—a testament to the fact that the GUI is merely a layer upon a powerful underlying system.

If we find that there are inadequacies in the GUI for us, we can edit these files by hand in order to get the functionality we need. However, there is one major caveat with this—we must ensure that we are attentive to the automatic settings produced by the GUI and ensure that any alterations we make are going to be compatible with the GUI, or else we risk breaking the interface entirely. This can be quite a hindrance if our system is to become complicated. In order to customize Asterisk further, Trixbox offers a tool called Configuration Editor, which will allow administrators to make changes directly in the CONF files. As mentioned earlier, you have to be careful what you edit, as the GUI can easily overwrite your settings. In order to make Trixbox more customizable, you will notice that there are configuration files labeled with `_custom.conf`; any edits within these files will be loaded into Asterisk and will not be overwritten by the GUI. For example, a common file used for most customizations is `extensions_custom.conf`.

A simple one-to-one PBX

Now that we have an overview of how the main features of the Trixbox/FreePBX system are customized, we can create a simple PBX for handling a single line and extension for a home user. We can also take the knowledge of call routing gained from previous chapters and apply it to Trixbox/FreePBX. All of the concepts remain the same, we just apply them differently, and the result is virtually indistinguishable.

Extensions

Firstly, we will configure our extensions by opening the **PBX Settings** section (Trixbox) or going to the FreePBX Dashboard. Click on **Setup** and find **Extensions** on the lefthand side. Then configure the extension screen as follows (you may wish to change some settings to fit your own needs):

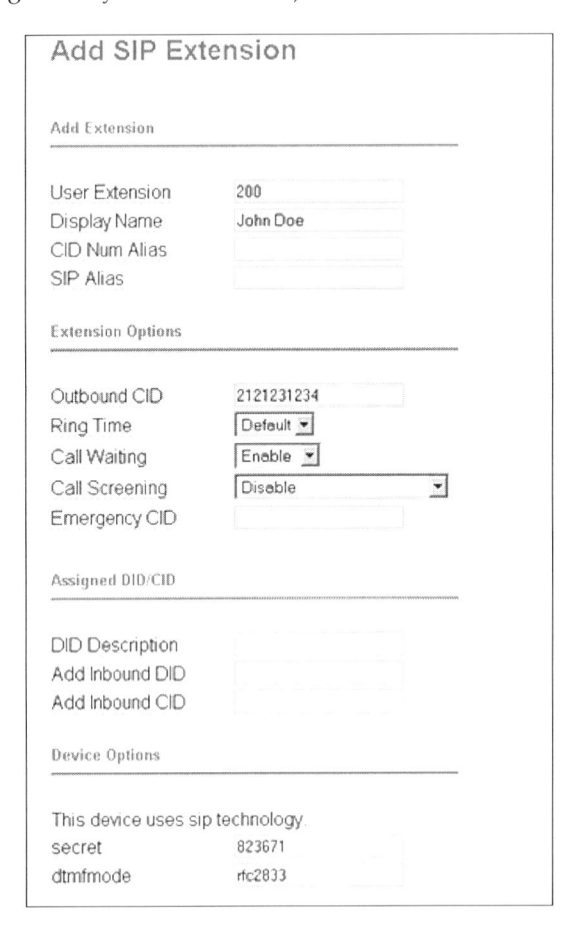

```
Add SIP Extension

Add Extension

User Extension      200
Display Name        John Doe
CID Num Alias
SIP Alias

Extension Options

Outbound CID        2121231234
Ring Time           Default
Call Waiting        Enable
Call Screening      Disable
Emergency CID

Assigned DID/CID

DID Description
Add Inbound DID
Add Inbound CID

Device Options

This device uses sip technology.
secret              823671
dtmfmode            rfc2833
```

After we have configured the extension, click on **Add Extension** on the bottom right hand corner of the screen. This sets up extension 200 for a SIP-based phone. We need to click on the red bar that appears afterwards, to apply the changes to the system.

Trunks

We can now add the trunk for our PSTN interface. We do so by clicking on **Trunks** on the lefthand side, clicking on **Add ZAP Trunk** (we can add other trunk types as discussed in previous chapters, such as SIP and IAX), and then configuring the trunk as follows (while here, we may also want to delete the default trunk g0):

In the **Outbound Caller ID** field, we would place our own phone number. Again remember to click on the red bar afterwards. It's important to note that when we make calls, there is often no check made against the caller ID number we present, so we could present anything here. We must verify that it's completely accurate or we may lose the ability for our contacts to recognize us and call us back. This can often be used to our advantage when we want to control the number we present. Also note that defining outbound caller ID here will force this caller ID on all extensions that use this Trunk. For those who might have different extensions requiring different caller IDs, it is best to leave this field blank and specify the outbound caller ID with the desired caller ID on an extension-by-extension basis.

Routes

Now that we have extensions and trunks, we require incoming and outgoing calling routes so that calls get to their correct destinations.

Firstly, create an incoming route by clicking on **Inbound Routes** and configuring it as follows:

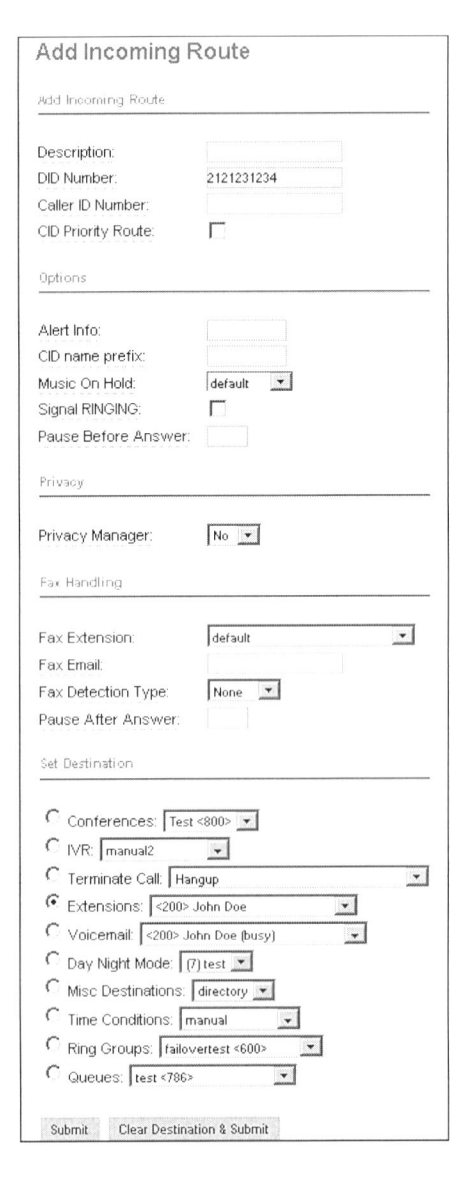

Then click on **Submit Changes** and the red bar to confirm.

We also require outgoing routes so that we can route our calls through the trunk that we have set up. Do this by clicking on **Outbound Routes** and then configuring it as shown in the screenshot that follows. You can modify the dial pattern here and can add alternative routes with differing patterns.

We should now be able to make and receive calls from our system over the PSTN. We should also have a working voicemail.

Summary

In this chapter, we have seen that Asterisk can easily be managed through a GUI. This makes it possible even for those with less technical skills to manage and make changes to Asterisk. It is important for those learning Asterisk to manually try installing Asterisk first and then manually add third-party applications. Trixbox is a great resource once you have mastered Asterisk on a stand-alone basis.

Another benefit with installing your own Asterisk build is the ability to host some of the services we have seen (such as FOP, ARI, and so on) on separate servers. This enables you to spread the load as well as the data. Therefore, if you expect significant growth or are starting with an already medium to large organization, you should consider building your Asterisk system from scratch.

8
What is asterCRM?

Now that we have covered installing a GUI to easily manage your Asterisk server, it's time to leverage the power of Asterisk with other business applications. A common business tool found in most businesses today is a **Customer Relationship Management (CRM)** system. Through this application, businesses can keep track of their customers, log activities, and streamline business operations. As most interactions with clients are often performed over the phone, it would be extremely useful to link CRM and Asterisk together.

An excellent open source application that bridges Asterisk and CRM together is asterCRM. asterCRM is an open source software application for call centers based on Asterisk. By connecting with Asterisk through the ami port using TCP protocol, asterCRM can work together with any system based on Asterisk. asterCRM, having adopted the advanced technology "AJAX", allows users to implement all the functions of a call center just by using a browser with a pop-up screen containing client information, Click to Dial, Call Record/Monitor, Speed Dial, and so on. Working with the basic needs of a call center, asterCRM provides CRM functions such as the management of user information, customer calling history, call recordings, and statistics management.

The application is actually combined with other applications in the form of a business suite called asterCC. The suite also packages asterBilling, which is a real-time billing solution for Asterisk. It could be used for a hosted callshop or to conduct Asterisk PBX billing. The suite is also available as an out of the box ISO solution similar to Trixbox, which installs and configures Asterisk, FreePBX, and asterCC automatically. asterCC manuals and software can all be downloaded from `http://astercc.org/downloads`.

What is asterCRM?

The following is a screenshot of the asterCRM interface:

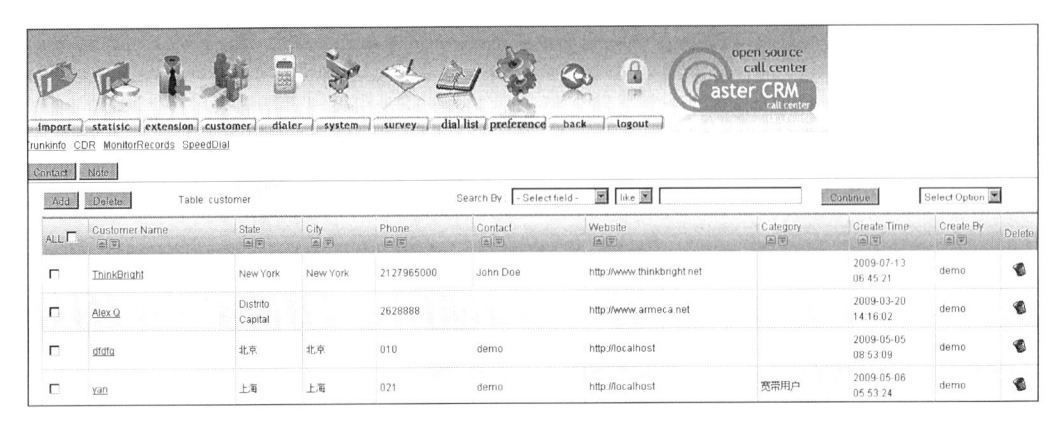

The following is a screenshot of the asterCRM interface with a pop-up display and customer call history:

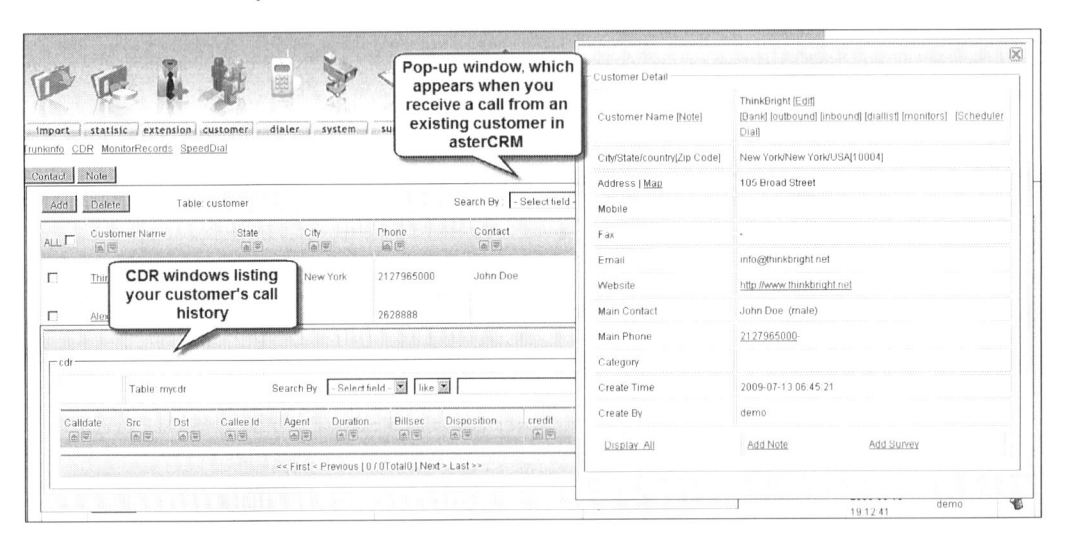

Installing asterCRM

The following is a list of instructions to install asterCC. Before you begin, please ensure you have the following installed on your Linux OS:

- httpd
- mysql
- mysql-devel

- mysql-server
- php (or php4)
- php-mysql
- php-gd

You can easily install the previous modules using `yum`.

Automatic installation

An auto-install script was created to make the installation quick and easy. In order to install asterCC using this nifty script just go to the main directory of the extracted asterCC and run the `install.sh` file.

Once completed, update the Asterisk dialplan for `barge`. If you want to use the `barge` function in asterCRM, add the following line to Asterisk's `extensions.conf` file. If you use any other extensions file in FreePBX, it could be `extensions_custom.conf`.

```
[astercc-barge]
exten => _X.,1,NoOP(${EXTEN})
exten => _X.,n,meetme(${EXTEN}|pqdx)
exten => _X.,n,hangup
```

Manual installation

Please follow steps 1 to 9:

1. **Download and unzip the source files**.

    Note that we will be assuming your WEB root is `/var/www/html`.

    We will begin by entering the following commands through the Linux command line:

    `#cd /var/www/html`

    `#wget http://astercc.org/download/astercc-0.12-beta.zip`

    `#unzip astercc-0.12-beta.zip`

    `#mv astercc-0.12-beta astercc`

    The explanation of each folder and file is listed as follows:

    ```
    /var/www/html/astercc/astercrm         # main directory and PHP
                                             scripts of astercrm
    /var/www/html/astercc/asterbilling     # main directory and PHP
                                             scripts of asterbilling
    ```

What is asterCRM?

```
/var/www/html/astercc/sql                    # sql to create database
                                               tables
/var/www/html/astercc/script                 # astercc daemon and some
                                               other script files
/var/www/html/astercc/index.html               #guide page
/var/www/html/astercc/astercc_full_logo.png    #logo
```

Please note that it is highly advised that the whole script directory be moved to a more secure location such as /opt and out of the WEB root directory.

2. **Create the MySQL database and tables**.

   Please note that we create the database named astercc, which is used for both asterCRM and asterBilling. asterCRM needs MySQL 4.1 or higher versions.

   You can use whatever database name you want, but be sure to use your configuration to replace yourmysqluser and yourmysqlpasswd.

   The following commands are to be entered through the Linux command line:

   `#mysqladmin -uyourmysqluser -pyourmysqlpasswd create astercc`

   `#mysql -uyourmysqluser -pyourmysqlpasswd astercc < /var/www/html/astercc/sql/astercc.sql`

3. **Update** /etc/asterisk/manager.conf **to enable manager connections**.

   In order to allow connections to Asterisk from a different server as well as to allow locally installed applications to communicate with Asterisk, you will need to add something like the following to the manager.conf file:

   > The following example is for those who installed asterCC on the same server as the Asterisk service.

```
enabled = yes
port = 5038
bindaddr = 0.0.0.0
;displayconnects = yes

;the following line could be changed by yourself
[astercc]
secret = astercc
read = system,call,log,verbose,command,agent,user
write = system,call,log,verbose,command,agent,user
deny=0.0.0.0/0.0.0.0
; if you want to run astercc on another server
; use your astercc ip to replace 127.0.0.1 or add a new line
permit=127.0.0.1/255.255.255.0
```

4. **Update the Asterisk dialplan for barge**.

   Follow this step only if you want to use the `barge` function in asterCRM. This will allow you to listen into a call that is currently in progress.

   Add following line to Asterisk's `extensions.conf` (for FreePBX, you would use `extensions_custom.conf`):

   ```
   [astercc-barge]
   exten => _X.,1,NoOP(${EXTEN})
   exten => _X.,n,meetme(${EXTEN}|pqdx)
   exten => _X.,n,hangup
   ```

5. **Create the directories and move the daemon scripts**:

   The following commands are to be entered through the Linux command line:

   ```
   #mkdir -p /opt/asterisk/scripts/astercc
   #mv /var/www/html/astercc/script/* /opt/asterisk/scripts/astercc
   #chmod +x /opt/asterisk/scripts/astercc/eventsdaemon.pl
   #chmod +x /opt/asterisk/scripts/astercc/eventdog.sh
   #chmod +x /opt/asterisk/scripts/astercc/astercc
   #chmod +x /opt/asterisk/scripts/astercc/astercctools
   #chmod +x /opt/asterisk/scripts/astercc/dialer.pl
   #chmod +x /opt/asterisk/scripts/astercc/cdr_move.pl
   #chmod +x /opt/asterisk/scripts/astercc/asterrc
   #chmod +x /opt/asterisk/scripts/astercc/astercclock
   #chmod +x /opt/asterisk/scripts/astercc/asterccdaemon
   #chmod +x /opt/asterisk/scripts/astercc/asterccd
   ```

6. **Modify the config file**.

   For asterCRM, modify `/var/www/html/astercc/astercrm/astercrm.conf.php` to fit your configuration.

   For asterCC, modify `/var/www/html/astercc/asterbilling/asterbilling.conf.php` to fit your configuration.

What is asterCRM?

7. **Start Asterisk and daemon**.

   There are two daemon modes you can choose: astercc mode or eventsdaemon (can be used for asterCRM only) mode.

   ○ **astercc mode (can be used for both asterCRM and asterBilling):**

   Try to start astercc:

   Modify /opt/asterisk/scripts/astercc/astercc.conf (mainly database setting and AMI setting) to fit your configuration.

   Run asterCC from /opt/asterisk/scripts/astercc/astercc for testing it.

   The following lines should appear on the Linux command line:

   Connecting to mysql database on 127.0.0.1:

   Database connection successful.

   Connecting to asterisk on 127.0.0.1 port 5038:

   Asterisk socket connection successful.

   Check asterisk username & secret:

   Success

   Monitor Start

   If you can read the previous lines, then congratulations! Your astercc works well. Press *Ctrl+C* to exit.

   If the previous lines are not displayed on the Linux command line and you get an error, then please check your database/AMI configuration in the astercc.conf file.

   Start astercc (default settings):

   Modify /var/www/html/astercrm/astercrm.conf.php; set event type to curcdr /opt/asterisk/scripts/astercc/astercc -d.

   Start up astercc daemons when the system boots up.

   Please note that this option only applies to Red Hat-release systems.

   If you want astercc daemons to start automatically when you boot your machine, then enter the following commands through the Linux command line:

   ```
   # cp /opt/asterisk/scripts/astercc/asterccd /etc/rc.d/init.d
   # chmod 755 /etc/rc.d/init.d/asterccd
   # chkconfig --add asterccd
   ```

It's advisable to configure your `astercc` to restart once everyday. It's not necessary, but it's good for the operation of your `astercc`. For example, if you want to restart `astercc` at midnight everyday, write the following line as root:

```
crontab -e
```

Add a new line:

```
0 0 * * * /etc/rc.d/init.d/asterccd restart
```

The first "0" denotes minutes and the second "0" denotes hours.

- **eventsdaemon mode (can be used for asterCRM only)**:

    Try to start `eventsdaemon`:

    Modify `/opt/asterisk/scripts/astercc/eventsdaemon.pl` (mainly database setting and AMI setting) to fit your configuration.

    The following line should appear on the Linux command line:

    ```
    Message: Authentication accepted
    ```

    If you can read the previous line, then congratulations! Your `eventsdaemon` works well. Press *Ctrl+C* to exit.

    If the previous line is not displayed on the Linux command line and you get an error, then please check your database/AMI configuration in `eventsdaemon.pl`.

    Start `eventsdaemon` (default settings):

    Modify `astercrm.conf`; set event type to event `/opt/asterisk/scripts/astercc/eventsdaemon.pl -d`.

    At some point, it may be desirable to delete unwanted events from the database table. The `eventsdaemon` mode is also designed for this. Please check `eventsdaemon.pl` for the `log_life` parameter.

    In order for `eventsdaemon` to be loaded everytime your server starts, add the following shell to your start-up file:

    ```
    # opt/asterisk/scripts/astercc/eventdog.sh >> /etc/rc.d/rc.local
    ```

8. **Set file and folder access**.

   For asterCRM:

   `# chmod 777 /var/www/html/astercc/astercrm/upload`

   `# chmod 777 /var/www/html/astercc/astercrm/astercrm.conf.php`

   If Asterisk and asterCRM are running on the same server, you can make a soft link to the asterCRM `web` directory for listening to monitored records online using the following command:

   `# ln -s /var/spool/asterisk/monitor/ /var/www/html/astercc/astercrm/monitor`

   For asterBilling:

   `# chmod 777 /var/www/html/astercc/asterbilling/upload`

9. **Web browsing—accessing asterCC using your web browser**.

   In order to access the asterCC guide, visit `http://YOUR-WEB-SERVER-ADDRESS/astercc`.

   For asterCRM,
   visit `http://YOUR-WEB-SERVER-ADDRESS/astercc/astercrm`.

   The default login is admin/admin.

   For asterBilling, visit
   `http://YOUR-WEB-SERVER-ADDRESS/astercc/asterbilling`.

   Please note that there are two login interfaces in asterBilling: `user` mode and `manager` mode. The default setting is `manager` mode.

   You can access the user interface through `http://YOUR-WEB-SERVER-ADDRESS/astercc/asterbilling/login.php`.

   You can access the manager interface through `http://YOUR-WEB-SERVER-ADDRESS/astercc/asterbilling/manager_login.php`.

   You can change the default login mode in `asterbilling.conf.php` by using the `useindex` parameter.

   You can also move the `astercrm` and `asterbilling` directories to any path where your web server is allowed access.

   Log in with admin/admin

   Set your first booth through asterBilling:

   1. Go to the "Reseller" section and add a reseller.
   2. Go to the "Group" section and add a group belonging to the reseller.

3. Go to the "Clid" section and add a CLID for this group. Then the account in Asterisk with a matching CLID will be billed as a user in this group.
4. Go to the "Account" section and add an account. The "usertype" could be "groupadmin" and belongs to the group you just added.
5. Go to the "Rate to Customer" section and add some rates for the group. If you don't select reseller or group, the rate will be the default rate which is applied to resellers/groups.
6. Log in under the group admin account.
7. Try making a call using the IP phone with the CLID. You should be able to see the calling and billing message in the box.

Introducing asterCRM

Now that we have installed asterCRM, we will go through the different sections of the application and explain their functions.

Import

In this section, you can import and upload client information from Excel or CVS files.

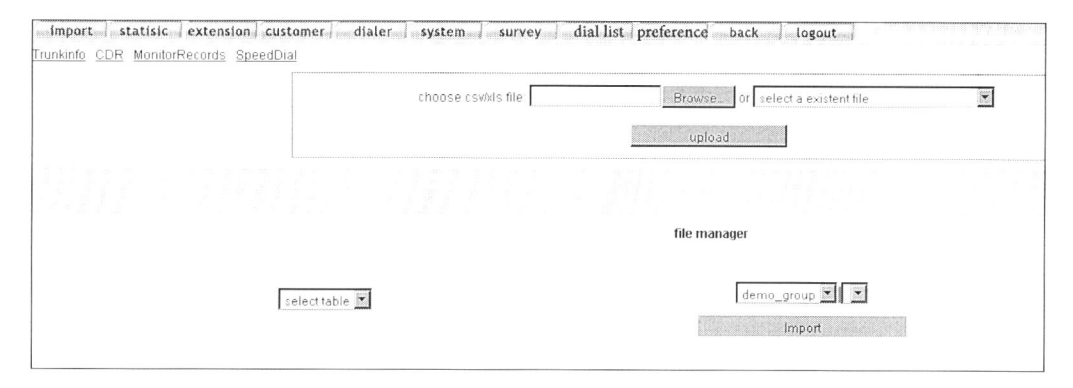

Statistic

Through the **statistic** section, you can pull up **Call Detail Records** (**CDR**s), or trunk information, as well as access calls that have been monitored.

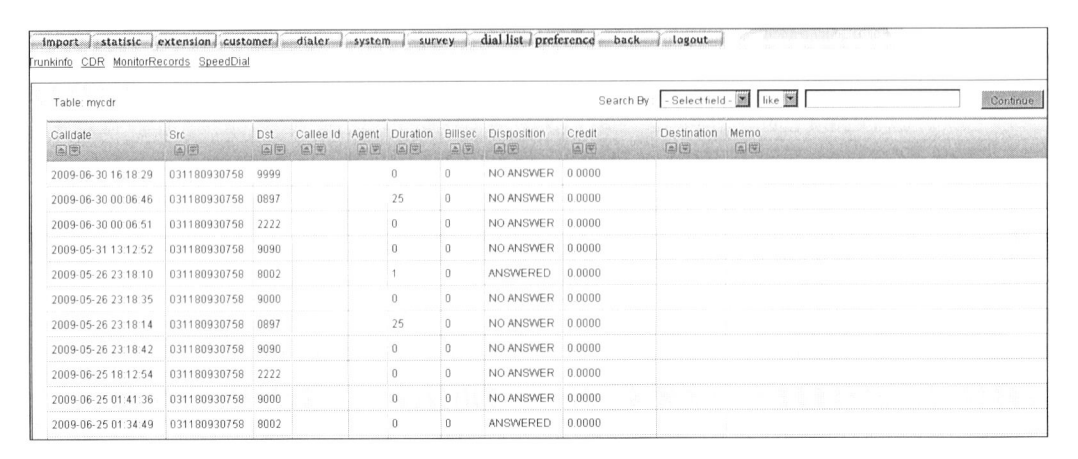

Extension

The **extension** tab is used to manage the users of the asterCRM system (the **astercrm_account** table). Here we can also bind an asterCRM account to an Asterisk extension. You can also add groups and users through this section.

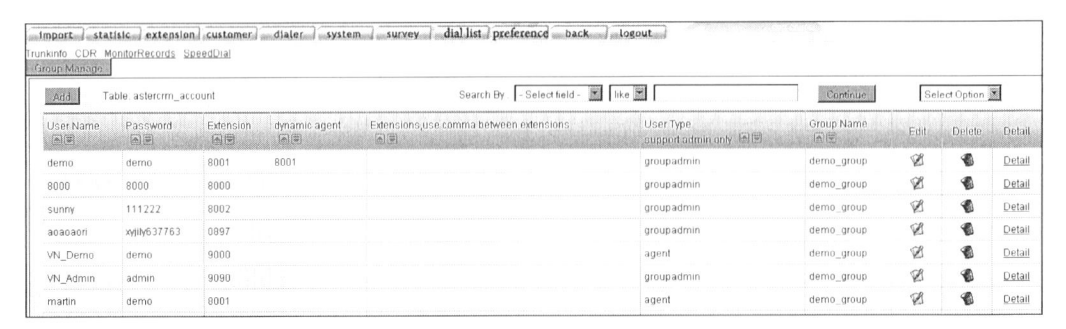

Customer

You can manage the information about a client by adding, deleting, modifying, and exporting data. You can also associate phone numbers with a client to allow a pop-up display to appear when your client calls. This section also gives you access to all calls placed to and from your client.

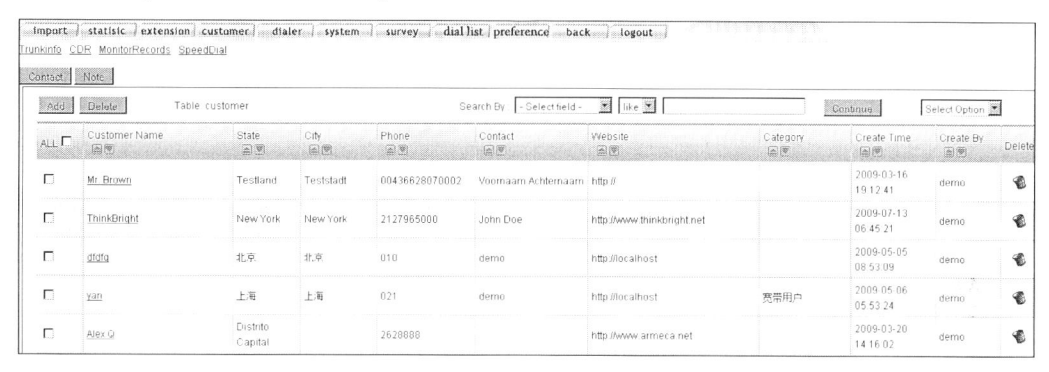

Dialer

The **dialer** section allows you to define settings for a predictive dialer. A predictive dialer can go through a list of numbers and when someone answers the phone, an IVR can be played to them. This can be a very powerful tool when trying to communicate a message or an advertisement campaign to all of your clients. Predictive dialers are often used to reach out to a large audience using an IVR. If the caller is interested and selects an option from the menu, then the call is transferred to an agent.

System

The **system** section allows you to see the call status of a given extension. You can also listen in to the call, hang up the call, or just see who is on the line.

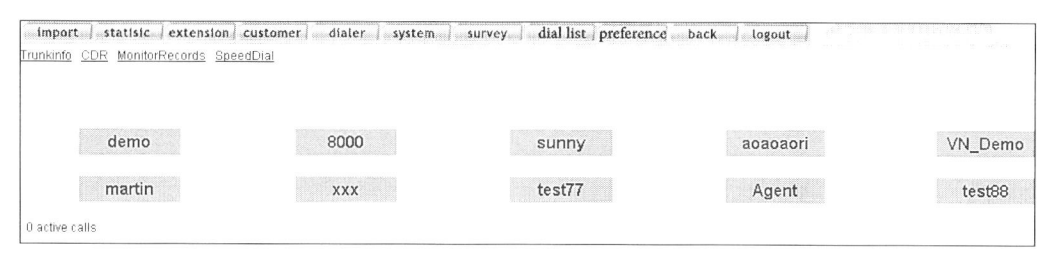

Survey

In this section, you can set up surveys that you can then use in marketing campaigns through your predictive dialer. This is an excellent way to collect data and information from your client base.

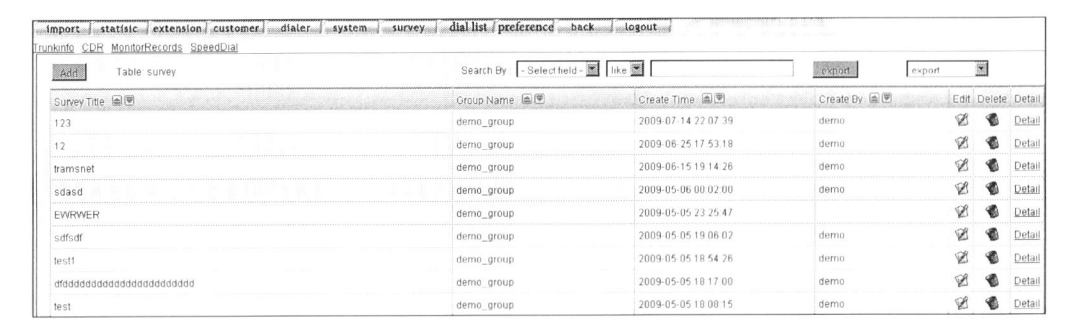

Summary

As we have seen, Asterisk can be used with third-party business applications such as CRM solutions. In this chapter, we used the asterCC suite. However, there are many other open source CRM applications that are equally popular. For example, SugarCRM (http://www.sugarcrm.com/crm) is another popular CRM application, which can be integrated with Asterisk. SugarCRM was also once bundled with Trixbox releases, but it needs to be downloaded separately today. If you search the Trixbox forums, you will find step-by-step instructions for installing SugarCRM with Trixbox. The main idea here is to understand that Asterisk is very adaptable and can be used with many types of business applications.

9
Case Studies

Up to this point, we have worked through the setting up of a new phone system based on Asterisk. Our system has been developed with our specific needs in mind and has configurations and features enabled to fit our particular purposes. Often we will find ourselves wondering if we're making full use of the magic of Asterisk, and perhaps the best way to find new tricks is to examine examples of working phone systems.

What follows in this chapter are typical examples of just a few such systems. Each section will be devoted to one type of system. First, we will give a brief overview of the type of customers involved before we mention some pointers to remember as we decide what they need and how best to accomplish our task. Finally, we will go through the configuration files, one at a time. Each configuration file will be annotated, and we will briefly discuss why some of the choices are particularly good for the given scenario.

Small office/home office

This is a common setup for Asterisk. In very small installations, Asterisk can be used to give us the features we need from an expensive PBX system at small office prices. Using Asterisk as a phone system for a home office gives the small business a big-business sound and feel.

The scenario

For our first example, we will join David in his home. An avid programmer and all-round geek, he's decided he wants to have his very own phone system in his house. As he is often busy and unable to answer the phone, the system must have voicemail. He has recently moved into an older home and only has one incoming line from the telephone company.

David has a new baby in the home and is very concerned about safety. He lives out in the middle of nowhere, and loses power pretty regularly. He only has phones in the office, kitchen, and master bedroom.

Finally, David would like to have music on hold for people who call in. As he is starting a new business, he wants to present a professional image to the callers he places on hold while finishing a call with another client or simply completing household tasks.

The discussion

First, Asterisk could be an appropriate choice here. As David is a programmer, he will be comfortable managing his own phone system. For home users who are not technically minded, Asterisk may not be a good fit, unless they are supported by a larger IT department at their employer.

Secondly, to meet David's requirements, we will have to use POTS lines. As he moved into an older house, running cable will be very difficult. Also, as he lives in an area prone to power failures, having the Asterisk server provide power to the telephones is a good thing, as this will require only one UPS.

Giving David voicemail and music on hold will be very simple. In fact, the thing to notice here, as with many SOHO configurations, is the lack of any other requests. This means we will not need to configure SIP, IAX, H.323, parking, menus, or any other advanced feature.

As he has three handsets and one incoming line, we will be using a Digium TDM31B card. This is a hardware TDM device that has three FXS ports and one FXO port. Also, as he has such low requirements, that is all calls will be directly bridged on TDM interfaces, he will be able to add Asterisk to his email server, which is already running Linux on a Dell PowerEdge 400 SC.

The configuration

Here are our configuration files for David. We should note that his server has been set up as we have previously discussed, including mpg123. Also, his house has home-run telephone wiring, meaning all telephone jacks terminate at the same closet.

Any files not listed here are left as default. If we did not install the default files, we may do so at any time by changing to the `/usr/src/asterisk` directory and executing `make samples`.

system.conf

The content of this file is as follows:

```
fxoks=1-3   ; 3 fxs modules, on channels 1-3
fxsks=4     ; 1 fxo module on channel 4
loadzone=us
defaultzone=us
```

We should remember that FXS modules use FXO signaling, and FXO modules use FXS signaling. Also, we should always put our FXS modules on lower port numbers because of some reported inconsistencies when putting an FXO module on channel 1 of a TDM card.

chan_dahdi.conf

The content of this file is as follows:

```
[trunkgroups]
[channels]
language=en
usecallerid=yes      ; He subscribes to CID
hidecallerid=no
callwaiting=no       ; but not call waiting...
usecallingpres=yes   ; NOTE: this does not always work right, but when
                     ;  it does, it is quite useful
transfer=yes
cancallforward=yes
callreturn=yes
echocancel=yes
echocancelwhenbridged=no
echotraining=800     ; he had echo until he set the train time to 800
; FXO Interface
context=default      ; all calls go to the "default" context
signalling=fxs_ks    ; we use FXS signaling for our FXO device
group=1              ; we are placing the outgoing line in a group
channel=>4

; FXS Interfaces
context=outgoing     ; All phones in the house may dial long distance
signalling=fxo_ks    ; we use FXO signaling for FXS devices
group=2              ; we are putting all internal phones in a group
pickupgroup=2        ; this is so we can pick calls from other lines
callgroup=2          ;  which may not be useful in this instance, but
                     ;  should not hurt anything
channel=>1-3         ; We select the channels
```

A few things we should notice are:

- We usually don't want to enable echo cancellation when calls are bridged. This can especially cause a problem with modem and fax communications. Also, we can modify the echo training period to the best value for our particular installation using a process of trial and error.
- We are not segregating the phones based on where they can call. It may be tempting to put all of the incoming and outgoing extensions in a single context; however, it is not wise to do so from a security standpoint. Thus, we have a blend of security and simplicity.

musiconhold.conf

The content of this file is as follows:

```
[classes]
default => quietmp3:/var/spool/asterisk/defaultMOH,-z
```

Here we have our music on hold configuration. Notice that we only have one class, which we called `default`. Also, we have chosen to shuffle our files so that the system doesn't always start with the same song. In small installations, it is very important to remember this because it is not very impressive if a customer calls frequently and, every time, they are greeted with exactly the same music on hold selections.

voicemail.conf

The content of this file is as follows:

```
[general]
format=wav49|gsm|wav
serveremail=asterisk@davidscomputer.com
attach=yes        ; voicemail messages will be attached to emails
skipms=3000
maxsilence=10
silencethreshold=128
maxlogins=6       ; David's wife isn't so good with passwords...

[zonemessages]
central=America/Chicago|'vm-received' Q 'digits/at' IMp

[default]
1 => 1234,David Gomillion,david@mydomain.tld,pager@mydomain.tld,tz=central
```

We have only one time zone, which is Central in the United States. We also have only one voicemail box. Asterisk can do a lot more, but in this instance, no more is needed.

The configuration choices we made at the beginning of the file are pretty much standard, except for the server's return email address. This should be set to something meaningful, if we are going to have users who will reply to these messages. However, in this instance, this is just a fake (but informative) address because David simply won't try to reply to these email notifications.

modules.conf

The content of this file is as follows:

```
[modules]
autoload=yes
noload => pbx_gtkconsole.so    ;don't load stuff we won't need
noload => pbx_kdeconsole.so
noload => chan_sip.so
noload => chan_iax.so
noload => chan_iax2.so
noload => chan_skinny.so
noload => chan_mgcp.so
noload => res_agi.so
noload => app_intercom.so
load => chan_modem.so
load => res_musiconhold.so

[global]
chan_modem.so=yes
```

In our `modules.conf` file, we have disabled all of the VoIP protocols that we will not be using. This will help increase the security of our server, as this keeps ports closed that have no need to be open. We have also firewalled all of the ports on the server except those needed for other servers.

extensions.conf

The content of this file is as follows:

```
[general]
static=yes
writeprotect=no

[globals]
TRUNK=Zap/g1
```

Case Studies

```
TRUNKMSD=1
[outgoing]
exten => _9.,1,Dial(${TRUNK}/${EXTEN:${TRUNKMSD}})  ;if we dial 9,
                                                    ; send to trunk
include => default
[default]
exten => s,1,Dial(Zap/g2,30)            ; dial all extensions for 30
                                        ; seconds
exten => s,2,Voicemail(u1)              ; send to VM if we don't pick up
exten => s,3,Hangup
exten => s,102,Voicemail(b1)            ; send to VM if we are busy
exten => s,103,Hangup

exten => 0,1,Dial(Zap/g2)               ; if we dial 0, ring all phones
exten => 1,1,Dial(Zap/1)                ; if we dial 1, ring the office
exten => 2,1,Dial(Zap/2)                ; if we dial 2, ring the bedroom
exten => 3,1,Dial(Zap/3)                ; if we dial 3, ring the kitchen
exten => 8,1,VoicemailMain(s1)          ; press 8 to check messages
                                        ;  without requiring password

exten => i,1,Goto(s,1)      ; if we are in an invalid or timed-out
exten => t,1,Goto(s,1)      ; state, go to s,1 in this context
```

This is our entire dialplan. We can see that it is very simple—each phone has an extension, and there is an extension for all phones. Only incoming calls are going to go to voicemail if a phone is busy or not answered.

We will notice that any number that is dialed with 9 as the first digit will automatically be sent out through the trunk. This is a very simple example of how a single pattern can accomplish many tasks. As we are not very concerned about securing the trunk from internal extensions, it is alright to use this simple method of trunk access.

Conclusions

As we can see, Asterisk configurations can be very simple. Creating a PBX for a few extensions is easy. Moreover, it illustrates some points that we will also see later in configuring some other PBX systems.

Small business

Small businesses make up a large portion of the IT market. These customers are unlike any others; they need upscale features with limited resources. It is very common for small businesses to require advanced features while needing to keep costs down. It's also common for small businesses to want to appear larger and more established than they are to increase customer confidence. Asterisk can be a great solution for small businesses as it suits these needs well.

The scenario

ACMEsoft is a software engineering firm with 40 employees. According to recent usage studies by their telephone company, they usually use about 18 lines, with their peak at 22 lines last month. They have a number of hosted extensions from the local telephone company (often referred to as Centrex service), which they have been using for years. As their five-year contract with the telephone company is up for renewal, they wish to replace the expensive hosted service with an in-house solution.

They will be contracting with us to provide the deployment, ongoing support, and maintenance of their new phone system.

ACMEsoft employs four first-tier support engineers, two second-tier support personnel, and one third-tier support specialist. Each member of each tier has similar talents and can handle the same calls.

They employ a receptionist, an operator, and an administrative assistant. There are 20 programmers, five testers, four project managers, and one person in the shipping department.

The discussion

Asterisk is an appropriate choice for ACMEsoft. Asterisk provides all of the features common to Centrex solutions. As they have no illusions of having an in-house tech to administer Asterisk, only our knowledge set is in question. As we are professionals who specialize in using Asterisk, we will be able to make it work according to ACMEsoft's expectations.

Asterisk is a powerful alternative to the more expensive hosted solutions. When using Centrex service, each extension must have an analog line. These lines are expensive to install, move, and maintain.

Case Studies

With the current usage statistics, a PRI line makes the most sense. The reason for this is that we will need less than 23 concurrent lines. PRI allows us to use advanced signaling; also, echo is less likely with a PRI than with POTS lines. PRI is often cheaper than having 23 separate POTS lines coming in to our server. Therefore, for this installation, PRI makes the most sense.

With Centrex service, each extension usually gets a unique phone number so that it may be reached from the outside world. In order to have the same feature, we will be using Directed Inward Dialing (DID) numbers. Usually purchased in blocks of 20, each number can be mapped to an extension, group of extensions, or a service, such as conferencing or voicemail. These numbers are generally inexpensive.

In this example, we will assume that the phone company will provide the full ten-digit phone number for each phone call. This is a very common configuration, which should be available from any phone company. We should always request the full ten digits in case we have the same last four digits for two telephone numbers coming into our system.

For our connection to the PSTN, we will be using Digium's T100P. This T1 card supports ISDN signaling and integrates well with Asterisk. For our handsets, we will be using Polycom's SoundPoint IP320 SIP phones.

The support personnel will be organized into queues; each level of support will have one queue. The operator will also have a queue, as he often receives multiple calls simultaneously.

The configuration

These are the configuration files for ACMEsoft's PBX. These files assume we have already set up our server as previously discussed.

system.conf

The content of this file is as follows:

```
#incoming PRI 1
span=1,1,0,esf,b8zs
bchan=1-23
dchan=24
loadzone = us
defaultzone=us
```

We are using ESF framing and B8ZS coding. In the United States, these are very common for PRIs.

chan_dahdi.conf

The content of this file is as follows:

```
switchtype=national
context=incoming
signalling=pri_cpe
group=1
channel => 1-23
```

Note that DAHDI and many other CPE devices refer to the "NI2" protocol as shown in the `switchtype` setting. If your PRI carrier tells you to use "NI2" for the protocol, do not put "NI2" as the value. Make sure its set to `national` as indicated above.

Here we are setting channels 1 through 23 (the channels that take actual calls; channel 24 is for signaling) to be in group 1, and we are telling incoming calls to go to the context called `incoming` in the dialplan.

musiconhold.conf

The content of this file is as follows:

```
[classes]
default => quietmp3:/var/spool/asterisk/defaultMOH,-z
```

Here we have a general music on hold instance, called `default`.

agents.conf

The content of this file is as follows:

```
[agents]
ackcall=yes
wrapuptime=0
musiconhold => default
updatecdr=yes
;Tier 1
group=1
agent => 1111,0596,John Smith
agent => 1209,0522,William Krandal
agent => 0186,1129,Lindsey Cramer
agent => 0416,0106,Stephanie Lewis
;Tier 2
group=2
agent => 2345234,3489,Likes Longnum
agent => 5692,4989,Smitty Rodriguez
```

```
;Tier 3
group=3
agent => 1,1,Forgets Ownname
;Operator
group=4
agent =>0,1234,Operator Console
```

Notice that we can have variable agent IDs. This is usually not a very good idea, as having consistent lengths for IDs is easier to support. However, politics will often dictate whether the length can be standardized.

queues.conf

The content of this file is as follows:

```
[general]

[default]

;
;Tier 1 Support Queue
[Q110]
music=default
strategy=leastrecent
maxlen=0
context=default
member => Agent/@1

;
;Tier 2 Support Queue
[Q120]
music=default
strategy=ringall
maxlen=0
context=default
member => Agent/@2

;
;Tier 3
[Q130]
music=default
strategy=leastrecent
maxlen=0
context=default
member => Agent/@3

;
;Operator Queue
```

```
[Q100]
music=default
strategy=ringall
maxlen=0
context=default
member => Agent/@4
```

Notice that each queue has its own section. We have configured each queue to have no limit as to length. We will be using some nifty options in the `extensions.conf` file to limit how long callers will be on hold, as setting the options upon entrance seems to be more reliable than setting them in the `queues.conf` file.

sip.conf

The content of this file is as follows:

```
[general]
context=default
port=5060
bindaddr=0.0.0.0
disallow=all
allow=ulaw

[101]
type=friend
context=local
callerid=ACMEsoft Operator<555-555-1234>
host=dynamic
secret=mypass101
dtmfmode=inband
mailbox=101

[102]
type=friend
context=longdistance
callerid=Sharon Stone<555-555-1235>
host=dynamic
secret=mypass102
dtmfmode=inband
mailbox=102

[111]
type=friend
context=default
callerid=John Smith<111>
host=dynamic
secret=mypass111
dtmfmode=inband
mailbox=111
```

As you can see, a clear pattern is emerging in this file. We simply copy and paste these configurations to create all 40 extensions needed. As we have all matching phones, we know that the DTMF mode will be the same for all of them. Also, as we are providing voicemail to all of our users, that will also be similar from user to user.

We should also take care to put our users in the proper context. Our first-level support agent can only call internal extensions, our operator can dial local and toll-free numbers, and our administrative assistant can dial long distance.

The rest of this example assumes that we have created the rest of the necessary entries; for the sake of brevity, they have been omitted here.

meetme.conf

The content of this file is as follows:

```
[rooms]
conf => 850
conf => 851
conf => 852
conf => 853
conf => 854
conf => 855
conf => 856
conf => 857
conf => 858
conf => 859
```

Here we have created ten conference rooms, with no passwords assigned.

voicemail.conf

The content of this file is as follows:

```
[general]
format=wav49|gsm|wav
serveremail=asterisk@mydomain.com
attach=yes
maxmessage=180
minmessage=3
maxgreet=60
skipms=3000
maxsilence=10
silencethreshold=128
maxlogins=1
```

```
fromstring=The Greatest PBX IN THE WORLD!!!

[zonemessages]
eastern=America/New_York|'vm-received' Q 'digits/at' IMp
central=America/Chicago|'vm-received' Q 'digits/at' Imp

[default]
100 => 100,Operator Queue Mailbox,,tz=central
101 => 123,Operators Mailbox,,tz=central
102 => 674,Patty Smalley,,tz=central
111 => 38594,John Smith,,tz=eastern    ;Support Department works by
                                       ; ETZ
112 => 65413,William Krandal,,tz=eastern
113 => 654,Lindsey Cramer,,tz=eastern
114 => 0106,Stephanie Lewis,,tz=eastern
```

As we can see, configuring voicemail is simple. The important thing to remember is that whatever we set the name to determines whether an extension will match an entry in the directory. Also, the context in voicemail should always match the context in `extensions.conf`.

extensions.conf

The content of this file is as follows:

```
[general]
static=yes
writeprotect=no

#include macros.incl
#include incoming.incl
#include outgoing.incl
#include default.incl
#include dialext.incl

[globals]
TRUNK=Zap/g1
TRUNKMSD=1
```

Case Studies

This is our entire extensions.conf file. By using the #include feature, we are able to make our configuration files much easier to read, and much easier to maintain. We should remember to keep the filenames easy to read and logical. As all of these files are included in the extensions.conf file, they will not get separate sections in this chapter.

```
;macros.incl
;#included into extensions.conf
[macro-stdexten]
;
; Standard extension macro:
;    ${ARG1} - Extension (we could have used ${MACRO_EXTEN} here as
;                         well)
;    ${ARG2} - Device(s) to ring
;
exten => s,1,Dial(${ARG2},20)             ; Ring the interface, 20
                                          ;   seconds maximum
exten => s,2,Goto(s-${DIALSTATUS},1)      ; Jump based on status

exten => s-NOANSWER,1,Voicemail(u${ARG1}) ; If unavailable, send to
                                          ;   voicemail
exten => s-NOANSWER,2,Goto(default,0,1)   ; If they press #, go to
                                          ;   Operator

exten => s-BUSY,1,Voicemail(b${ARG1})     ; If busy, send to voicemail
                                          ;   with busy message
exten => s-BUSY,2,Goto(default,0,1)       ; If they press #, go to
                                          ;   Operator

exten => s-CHANUNAVAIL,1,Voicemail(u${ARG1})
exten => s-CHANUNAVAIL,2,Goto(default,0,1)

exten => s-.,1,Goto(s-NOANSWER,1)         ; Treat anything else as no
                                          ;   answer
exten => a,1,VoicemailMain(${ARG1})       ; If they press *, send to
                                          ;   VoicemailMain

[macro-novm]
exten => s,1,Dial(${ARG1},30)        ;ring the interface for 30 seconds
exten => s,2,Goto(default,s,1)
exten => s,102,Goto(default,s,1)
```

Notice that we have a macro to set up all of the extensions we will be creating. This will save us a ton of work later on, as well as make our configuration files very readable.

```
;incoming.incl
;#included from extensions.conf
[incoming]
exten => 5555551234,1,Goto(default,100,1)  ;Main number rings to
                                           ;  Operators
exten => 5555552345,1,Goto(default,110,1)  ;Direct number to Support
exten => 5551110001,1,Goto(default,111,1)  ;Direct line to
                                           ;  Extension 111
exten => 5551110002,1,Goto(default,112,1)  ;Direct line to
                                           ;  Extension 112
exten => 5551110003,1,Goto(default,113,1)  ;Direct line to
                                           ;  Extension 113
. . .
exten => s,1,Goto(default,100,1);
exten => t,1,Goto(default,100,1);
exten => i,1,Goto(default,100,1);
```

Notice that we handle all incoming calls via this file. Here we define our DIDs and where we want them to ring. We also make sure to create intelligent rules in case the DID information is mangled by our phone company before Asterisk can decode it. In this case, we are sending the calls to our operator.

```
; outgoing.incl
;#included from extensions.conf
[local]
ignorepat => 9
exten => _9NXXXXXX,1,Goto(trunkdial,${EXTEN},1)
exten => _91800XXXXXXX,1,Goto(trunkdial,${EXTEN},1)
exten => _91866XXXXXXX,1,Goto(trunkdial,${EXTEN},1)
exten => _91877XXXXXXX,1,Goto(trunkdial,${EXTEN},1)
exten => _91888XXXXXXX,1,Goto(trunkdial,${EXTEN},1)
include => default

[longdistance]
ignorepat => 9
exten => _91NXXNXXXXXX,1,Goto(trunkdial,${EXTEN},1)
include => local

[trunkdial]
exten => _9.,1,Dial(${TRUNK}/${EXTEN:${TRUNKMSD}})
exten => _9.,2,Congestion(5)
exten => _9.,3,Hangup
```

Case Studies

Notice what we have done here—we created a general context called `trunkdial`, which we use to dial any calls going over the trunk lines. Why is this helpful? If we were to add a new trunk group, we could add only one line. If we were to use the standard method of having each line above dial, then we would have to add six lines for each new trunk group.

This example assumes we will have no users placed directly in the `trunkdial` context, such as in the `sip.conf` file. For security reasons, we must be careful that we do not ever place a user explicitly in the `trunkdial` context.

```
;default.incl
;#included in extensions.conf
[default]
exten => s,1,Goto(default,100,1)
exten => t,1,Goto(default,100,1)
exten => i,1,Goto(default,100,1)

; Operator queue, Operator Console, and Receptionist Phone
exten => 100,1,Answer
exten => 100,2,Queue(Q100||||240)   ; only allow 4 minutes in queue
exten => 100,3,Voicemail(u100)      ;   then send to VM
exten => _10[12],1,Macro(stdexten,${EXTEN},SIP/${EXTEN})

;Support Tier 1
exten => 110,1,Answer
exten => 110,2,Queue(Q110||||240)   ; allow 4 minutes in queue
exten => 110,3,Goto(default,100,1)  ;   then send to Operator
exten => _11[1-4],1,Macro(stdexten,${EXTEN},SIP/${EXTEN})


;Support Tier 2
exten => 120,1,Answer
exten => 120,2,Queue(Q120||||240)   ; allow 4 minutes in queue
exten => 120,3,Goto(default,100,1)  ;   then send to Operator
exten => _12[12],1,Macro(stdexten,${EXTEN},SIP/${EXTEN})

;Support Tier 3
exten => 130,1,Answer
exten => 130,2,Queue(Q130||||240)   ; allow 4 minutes in queue
exten => 130,3,Goto(default,100,1)  ;   then send to Operator
exten => 131,1,Macro(stdexten,${EXTEN},SIP/${EXTEN})

;Programmers, extensions 200-219
exten => _2[01]X,1,Macro(stdexten,${EXTEN},SIP/${EXTEN})
```

```
;Testers, extensions 251-255
exten => _25[1-5],1,Macro(stdexten,${EXTEN},SIP/${EXTEN})

;Project Managers, exts 301-304
exten => _30[1-4],1,Macro(stdexten,${EXTEN},SIP/${EXTEN})

;Shipping Department, ext 191, doesn't need voicemail
exten => 191,1,Macro(novm,SIP/${EXTEN})

exten => 800,1,Answer
exten => 800,2,VoicemailMain

exten => _85X,1,Answer
exten => _85X,2,MeetMe(${EXTEN})

exten => 888,1,Goto(dialext,s,1)
```

Notice that we are able to create 20 extensions for our programmers in a single line. This is the power of Asterisk's pattern matching, coupled with the flexibility of macros. We can use tricks like this one by grouping similar extensions together.

```
;dialext.incl
;#included from extensions.conf
[dialext]
include => default
exten => s,1,Answer
exten => s,2,DigitTimeout(5)
exten => s,3,ResponseTimeout(20)
exten => s,4,Background(pleaseenterextension) ; "Please enter the
                                 ;  extension of the party you are calling."

exten => 9,1,Directory(default)           ; press 9 for the directory
exten => 9,2,Goto(dialext,9,1)

exten => 0,1,Goto(default,100,1)          ; send to operator as a
                                          ;  courtesy if they press 0

exten => i,1,Playback(privacy-incorrect)
exten => i,2,Goto(dialext,s,1)

exten => t,1,Goto(dialext,i,1)
```

This small context allows users to dial anybody in the company, and also to access the corporate directory. The directory that reads the `voicemail.conf` file allows access to any extension in the company. By doing so, a "backdoor" line can be established that points directly to this extension, allowing us to no longer need direct phone numbers for each extension.

Conclusions

Asterisk has proved itself again as a powerful solution for real-world problems. By taking advantage of the feature set that Asterisk provides, we are able to create a server that has the features of a Centrex system. The savings that ACMEsoft will experience are very real and will pay for the system with a short ROI.

Hosted PBX

Asterisk is not limited to being able to service only one company. With a little finesse, we can configure Asterisk to handle multiple companies without the need to be aware of the others' presence. Although our example will deal with multiple companies on one site, there is no reason the same principles could not be applied over a high-speed data network.

The scenario

Al's Computer Depot was a very large computer retailer in the early 1990s, back when computers were fun and profitable. Unfortunately, Al got a bit too used to very high margins on computer sales, and has moved out of the computer selling business. He and his team have moved into consulting. As a consulting firm, 90% of the employees are traveling at any time.

Al's wife operates a small boutique selling Asian knock-off wallets. As most of the offices are empty all of the time, Al decided to let her have an office to run her business from. And with such a business, Sue needed a telephone line, but it would be no good to have the line answered by someone from Al's Computer Depot (the name wasn't changed as Al didn't want to buy new stationary when he went into consulting).

After this experience, Al decided to sublease more offices, requiring the tenants to purchase the telephone service from him. And so, Al's phone system is configured to allow multiple businesses on the same server.

We will be considering three separate businesses as follows:

- Al's Computer Depot
- Sue's Collectibles
- AutoAuction Listings

The discussion

Asterisk is perfect for this scenario. The flexible feature set will allow enough features for potential tenants, while being able to be scaled down for the smaller businesses who only need one line, like Sue's Collectibles.

We will be using SIP hardphones and a PRI line for PSTN interconnection. This will give us flexibility to change quickly when tenants come and go.

The configuration

Once again assuming we have properly installed Asterisk, the following files will configure our server for Al.

system.conf

The content of this file is as follows:

```
#incoming PRI 1
span=1,1,0,esf,b8zs
bchan=1-23
dchan=24
loadzone = us
defaultzone=us
```

Again, we use ESF framing and B8ZS coding, as they are very common for PRIs in the United States.

chan_dahdi.conf

The content of this file is as follows:

```
switchtype=national
context=incoming
signalling=pri_cpe
group=1
channel => 1-23
```

Case Studies

Here we are setting channels 1 through 23 (the channels that take actual calls; channel 24 is for signaling) to be in group 1, and we are telling incoming calls to go to the context called incoming in the dialplan.

musiconhold.conf

The content of this file is as follows:

```
[classes]
default => quietmp3:/var/spool/asterisk/defaultMOH,-z
```

Here we have a general music on hold instance, called default.

sip.conf

The content of this file is as follows:

```
[general]
context=default
port=5060
bindaddr=0.0.0.0
disallow=all
allow=ulaw

[al100]
type=friend
context=al-ld
callerid=Al Getrich<800-555-1234>
host=dynamic
secret=badpassword
dtmfmode=inband
mailbox=100@al
accountcode=al
 . . .
[sue1]
type=friend
context=sue-ld
callerid=Sue Getrich<555-555-5555>
host=dynamic
secret=anotherbadpassword
dtmfmode=inband
mailbox=1@sue
accountcode=sue

[aa100]
```

```
type=friend
context=aa-ld
callerid=Auto Auctions<555-777-1234>
host=dynamic
secret=1234
dtmfmode=inband
mailbox=100@aa
accountcode=aa
```

Here we see three of the many SIP extensions that are defined. Notice that two of the users, namely `aai100` and `al100`, both have mailboxes of `100`, and both will be extension `100`. As they are in different contexts, though, this will not be a problem.

Also, we should be sure to put each of the SIP users into the correct account codes. By doing so, we can correctly bill calls made to the party who made them. There are many billing solutions available for Asterisk; however, we focus on what can be done with the default setup here.

voicemail.conf

The content of this file is as follows:

```
[general]
format=wav49|gsm|wav
serveremail=asterisk@mydomain.com
attach=yes
maxmessage=180
minmessage=3
maxgreet=60
skipms=3000
maxsilence=10
silencethreshold=128
maxlogins=1
fromstring=The Greatest PBX IN THE WORLD!!!

[zonemessages]
central=America/Chicago|'vm-received' Q 'digits/at' IMp

[al]
100 => 100,Al Getrich,,tz=central

. . .

[sue]
1 => 1,Sue Getrich,,tz=central

[aa]
100 => 1234,Auto Auctions,,tz=central
```

Case Studies

This is a sample of how the `voicemail.conf` file will look, with one sample from each company. As we can see, configuring voicemail is simple. The important thing to remember is that whatever we set the name to determines when an extension will match an entry in the directory. Also, the context in voicemail should always match the context in `extensions.conf`. This allows each company to have its own directory, if it so chooses.

extensions.conf

The content of this file is as follows:

```
[general]
static=yes
writeprotect=no

#include macros.incl
#include al.incl
#include sue.incl
#include aa.incl
#include outgoing.incl
[globals]
TRUNK=Zap/g1
TRUNKMSD=1
```

This is our entire `extensions.conf` file. By using the `#include` feature, we are able to make our configuration files much easier to read and maintain. We should remember to keep the filenames easy to read and logical. As all of these files are included in the `extensions.conf` file, they are not given separate sections in this chapter.

```
;macros.incl
;#included into extensions.conf
[macro-stdexten]
;
; Standard extension macro:
;    ${ARG1} - Extension (we could have used ${MACRO_EXTEN} here as
;                         well)
;    ${ARG2} - Device(s) to ring
;
exten => s,1,Dial(${ARG2},20)          ; Ring the interface, 20 seconds
                                       ;   maximum
exten => s,2,Goto(s-${DIALSTATUS},1)   ; Jump based on status

exten => s-NOANSWER,1,Voicemail(u${ARG1}); If unavailable, send to
                                         ;   voicemail
```

```
exten => s-NOANSWER,2,Goto(default,0,1)     ; If they press #, go to
                                            ;   Operator

exten => s-BUSY,1,Voicemail(b${ARG1})       ; If busy, send to
                                            ;   voicemail with busy
exten => s-BUSY,2,Goto(default,0,1)         ; If they press #, go to
                                            ;   Operator

exten => s-CHANUNAVAIL,1,Voicemail(u${ARG1})
exten => s-CHANUNAVAIL,2,Goto(default,0,1)

exten => s-.,1,Goto(s-NOANSWER,1)           ; Treat anything else as
                                            ;   no answer

exten => a,1,VoicemailMain(${ARG1})         ; If they press *, send to
                                            ;   VoicemailMain

[macro-novm]
exten => s,1,Dial(${ARG1},30)               ; ring the interface for 30
                                            ;   seconds
exten => s,2,Goto(default,s,1)
exten => s,102,Goto(default,s,1)
```

We can actually reuse these macros from the previous example; this is why macros are so powerful. By defining things generically enough, we are able to reuse the same configuration in many different scenarios.

```
;al.incl
;#included from extensions.conf
[al]
exten => 8005551234,1,Goto(al,100,1) ;AL's direct number
exten => 100,1,Macro(stdexten,100@al,SIP/al100) ;AL's extension
```

Notice that we can handle incoming calls and internally dialed extensions. Here we define our DIDs and where we want them to ring. We also define failover behavior in case of bad information from our phone company. Finally, we define the extensions for Al's own business directly in this file.

One somewhat interesting side effect of this method is that if we dial the full ten-digit number from a telephone, it will route it internally, instead of hopping off to the PSTN and then coming back in. Of course, if we dial a *9*, then we will still use the trunk rules.

```
;sue.incl
;#included from extensions.conf
[sue]
exten => 5555555555,1,Goto(sue,1,1) ; Sue's direct number
```

Case Studies

```
exten => 1,1,Macro(stdexten,100@sue,SIP/sue1) ; Sue's extension
 . . .

;aa.incl
;$included from extensions.conf
[aa]
exten => 8005551234,1,Goto(al,100,1) ; Only phone number for A A
exten => 100,1,Macro(stdexten,100@aa,SIP/al100) ; Only AA extension
```

Here we have the other two businesses. We have chosen to configure them in much the same way as we did for Al. Each business will be in its own configuration file.

```
;outgoing.incl
;#included from extensions.conf
[al-local]
ignorepat => 9
exten => _9NXXXXXX,1,Goto(trunkdial,${EXTEN},1)
exten => _91800XXXXXXX,1,Goto(trunkdial,${EXTEN},1)
exten => _91866XXXXXXX,1,Goto(trunkdial,${EXTEN},1)
exten => _91877XXXXXXX,1,Goto(trunkdial,${EXTEN},1)
exten => _91888XXXXXXX,1,Goto(trunkdial,${EXTEN},1)
include => al

[al-ld]
ignorepad => 9
exten => _91NXXNXXXXXX,1,Goto(trunkdial,${EXTEN},1)
include => al-local

[sue-local]
ignorepat => 9
exten => _9NXXXXXX,1,Goto(trunkdial,${EXTEN},1)
exten => _91800XXXXXXX,1,Goto(trunkdial,${EXTEN},1)
exten => _91866XXXXXXX,1,Goto(trunkdial,${EXTEN},1)
exten => _91877XXXXXXX,1,Goto(trunkdial,${EXTEN},1)
exten => _91888XXXXXXX,1,Goto(trunkdial,${EXTEN},1)
include => sue

[sue=ld]
ignorepad => 9
exten => _91NXXNXXXXXX,1,Goto(trunkdial,${EXTEN},1)
include => sue-local

[aa-local]
ignorepat => 9
exten => _9NXXXXXX,1,Goto(trunkdial,${EXTEN},1)
exten => _91800XXXXXXX,1,Goto(trunkdial,${EXTEN},1)
```

```
exten => _91866XXXXXXX,1,Goto(trunkdial,${EXTEN},1)
exten => _91877XXXXXXX,1,Goto(trunkdial,${EXTEN},1)
exten => _91888XXXXXXX,1,Goto(trunkdial,${EXTEN},1)
include => aa

[aa-ld]
ignorepad => 9
exten => _91NXXNXXXXXX,1,Goto(trunkdial,${EXTEN},1)
include => aa-local

[trunkdial]
exten => _9.,1,Dial(${TRUNK}/${EXTEN:${TRUNKMSD}})
exten => _9.,2,Congestion(5)
exten => _9.,3,Hangup
```

Here we used the same `trunkdial` context as in the previous example for exactly the same reasons. However, with all these different outgoing contexts, the complexity of adding extensions would simply keep increasing.

Each company has a unique set of outgoing contexts so that only its extensions are included. This helps to ensure that the correct extension is reached when extensions exist in more than one context, such as extension 100 existing in contexts `aa` and `al`.

Compared to our previous example, this system offers fewer features for our users. We do not have conferences, we cannot get to the directory, and we must call our own extension and press the * key to get to our voicemail. However, these features can be activated at will, or even as customers pay to have them added.

Conclusions

Asterisk has again been used to fulfill a different need in a phone system. By taking advantage of contexts, we have been able to create multiple virtual phone servers with only one server, one PRI line, and one set of configuration files.

Summary

After looking at these case studies, I hope you have been able to spot common features in what we are trying to implement in each different situation. We can learn from these implementations and apply many of the same strategies when we encounter users with similar needs.

10
Maintenance and Security

Now that we have an Asterisk server installed and running, we should consider the maintenance and security of the server. There are a number of aspects involved here and we will cover each in turn. As the Asterisk server is going to be the central hub of our phone system, the importance of securing and maintaining it is obvious, as without it we lose our primary means of communication.

This chapter also looks at scalability issues, which are important to keep Asterisk running at high loads. Finally, the last section takes a look at how to get support from the community, and how to stay abreast of Asterisk developments. Keeping up with what's going on in the Asterisk world can be a very useful way to stay prepared for potential problems before they happen, as well as providing a helping hand should things go wrong.

Backup and system maintenance

One of the most important aspects of maintaining a system is the update or patch management process. It's vital that we keep our system up to date in order to reduce bugs and ensure that any security vulnerabilities in our software are fixed as soon as possible. When updating our Asterisk server, there are three main areas to maintain:

- The Asterisk service itself
- The various components that Asterisk depends on (DAHDI, LibPRI, festival, and so on)
- The host OS and any supplementary tools installed (OpenSSH, mpg123, and so on)

When we discussed setting up the Asterisk server, we covered installing from the CVS source repository, which is an easy way to keep Asterisk up to date, as you can continually download the latest version.

If we were to manage all applications as source packages only, we would need to go through the steps given in the installation chapter for each component as updates are released. However, when we factor in the updates for the tools of the host operating system and any other tools we use, this can become tedious and error prone quite quickly as these extras aren't contained in the Asterisk CVS repository. It is at this point that we should consider a package management system to ensure we keep everything up to date automatically, reducing our administrative burden. The options we have for this are dependent on the Linux distribution we decided to use.

Examples are:

- APT: Used by Debian (and ported to many other distributions such as Red Hat). Visit `http://www.debian.org/doc/manuals/apt-howto/index.en.html`.
- Portage: Used by Gentoo. Visit `http://www.gentoo.org/doc/en/handbook/handbook-x86.xml?part=2&chap=1`.
- URPMI: Used by Mandrake. Visit `http://www.urpmi.org`.
- Yum: Used by YellowDog/CentOS. Visit `http://linux.duke.edu/projects/yum/`. `Yum` is the command you will use in Trixbox installations.

Each of these is documented on its respective site. In order to ensure our system remains well maintained, we must become familiar with the package management tools at our disposal.

Backing up configurations

In the installation chapter, we briefly touched on backups and mentioned making a copy of the configuration files before editing them. This is good administrative practice and will protect us from any mistakes we make in the configuration files. However, it will not protect our configuration if the system were to be compromised or if we were to lose the media containing the configuration. It also doesn't take into account any data the system holds that we might lose.

Asterisk configurations can become quite involved over time and we invest a lot of time in setting these up. Repeating this work would be far from desirable in the event of system failure. For this reason, we should keep offline copies of the configuration files.

The backup method is not important. What is important is that we understand it and are sure that it fits our need. In the end, the final choice will mostly be down to personal preference. We have chosen some of the most common methods to provide practical examples. We can choose any backup solution that suits us or fits our business requirements as long as we ensure that we back up **configuration files**, **data**, and **logs**. The locations of these files are configurable and may change depending on our distribution. You can find their locations in the /etc/asterisk/asterisk.conf file:

```
[global]
astetcdir => /etc/asterisk
astmoddir => /usr/lib/asterisk/modules
astvarlibdir => /var/lib/asterisk
astagidir => /usr/share/asterisk/agi-bin
astspooldir => /var/spool/asterisk
astrundir => /var/run/asterisk
astlogdir => /var/log/asterisk
```

A simple manual copy of the /etc/asterisk directory may seem to be enough to ensure we don't lose these settings. However, we have to consider files outside the Asterisk directory that the server directly relies on, such as /etc/dahdi/system.conf and our network configuration files. The locations of some of the files important to us can vary with distribution, but invariably they are kept as a subfolder of the /etc/ directory. This means that as long as we ensure that this directory is copied from the server periodically, we can restore the configuration files in the event of a failure.

We should ensure this is done automatically at least once per week. There are a number of options for doing so but the simplest and easiest to automate is an rsync copy to our backup server, or if the Asterisk server has a tape drive, we could use the tar utility.

> **Backup schedule**
> Our backup schedule may have to be more frequent than once per week depending on business and personnel requirements.

Below is a sample rsyncd.conf configuration file:

```
[asterisk_backup]
    path = /home/adminuser/asterisk_backups
    comment = Asterisk Backups
    uid = nobody
    gid = nobody
```

Maintenance and Security

```
        read only = no
        list = no
        auth users = username
        secrets file = /etc/rsyncd/asterisk
```

The `/etc/rsyncd/asterisk` file would contain our username and password pair. This and the `/etc/rsyncd.conf` file would be on the backup server, which we would also install and run `rsyncd` on.

We could then run the following command on the Asterisk box in order to back up the configurations:

```
$ rsync --verbose --compress --progress --recursive --times --perms
   --nodelete /etc/* backupserver:asterisk_backup
```

This would copy everything from the `/etc/` directory on the Asterisk server to the `/home/adminuser/asterisk_backups` directory on the backup server.

> **Added security**
>
> If we install SSH on the backup server, then we could run the previous command over an encrypted SSH tunnel, by altering the command as follows:
>
> ```
> $ rsync --rsh=/usr/local/bin/ssh --verbose --compress -
> progress
>     --recursive --times --perms --nodelete /etc/*
>    backupserver:asterisk_backup
> ```

We could also back up to a tape drive attached to the Asterisk server by using the `tar` utility, which requires no configuration files as it is completely command-line driven:

```
$ tar --verbose -j --create /etc/*
```

The `-j` option compresses the files and subfolders under `/etc/` using `bzip2` for efficient storage.

If we want to use an archive file instead of a tape drive, we add the `--file` (or `-f`) parameter:

```
$ tar --file=asterisk_backup.tar.bz2 --verbose -j --create /etc/*
```

We could automate these processes by adding the relevant command lines to cron. For example:

```
$ crontab -e
 0 2 * * 0 tar --file=asterisk_backup.tar.bz2 --verbose -j
          --create /etc/*
```

This would back up the configuration files at 02:00 every Sunday morning. It may be worth studying cron if we are unfamiliar with it to ensure we understand its capabilities.

Backing up voice data

We would also want to ensure we had backups of the saved data that we have in our spool directory. This is where our hold music and voicemail is located. We would almost certainly want to restore this in the event of a system failure, especially when we have referenced audio files in our configuration files and hence, they are required for Asterisk to function properly. The default location of the spool directory is `/var/spool/asterisk`. Hence, we should add it to our backup commands.

Using `rsync`:

```
$ rsync --verbose --compress --progress --recursive --times –perms
    --nodelete /etc/* /var/spool/asterisk backupserver:asterisk_backup
```

Using `tar`:

```
$ tar --file=asterisk_backup.tar.bz2 --verbose -j --create /etc/*
    /var/spool/asterisk
```

This would ensure our backups contain all of our configuration and voice data that is necessary for a complete restoration of the Asterisk server.

> **Additional considerations**
> We may also want to back up other data and log files if we have additional applications installed with the Asterisk server, such as Webmin.

Backing up log files

It is important to ensure that any logs we have are backed up too, and we can do this by adding `/var/log/asterisk` to the commands. It may not always be necessary to keep log files. However, we may have to keep these if we have a data retention policy or are under regulations that require log and data to be kept.

> **Policies**
> If we are under a regulation or policy such as Sarbanes-Oxley or HIPAA, then we will have clearly defined rules for log retention and may have to ensure compliance of the Asterisk backups.

Backup scripts

In order to back up the system effectively, it's important that we back up the system incrementally so that we can go back to previous points in time, for example, before a disaster happened or before a configuration change that caused problems. The scripts that follow allow us to do so by keeping copies of our configurations and data on a backup server, and running these daily means we can go back and restore any date's configuration if necessary.

The following scripts allow us to back up our files to a backup server running an SSH daemon. They can be used as is (with the obvious change to the `backup_server` variable) or as a starting point for a backup system. The files are commented enough to make them self-explanatory. They use a combination of `tar` and `scp` to archive the files and copy them off the server. Alternatively, you could use `rsync` by following the guidelines shown earlier.

First, here is the `backup.cron` script:

```
#!/bin/bash

#########################################
## Backup script for asterisk          ##
#########################################
## This script is designed to make     ##
## a copy of all important config      ##
## and data files, which will then     ##
## be copied to a backup server        ##
## running an ssh daemon               ##
#########################################
## Usage: backup.cron, no              ##
## arguments required.                 ##
#########################################


# edit variable below to contain your
# backup server user and hostname as
# well as directory location
backup_server="username@backupserver:/path/to/backups"
date=`date +%Y-%m-%d`

# Remove old backups to keep from filling the disk with junk
rm /backup/*.tar.gz -f

# Backup the /etc/ directory
tar cfz /backup/asterisk-configs-${date}.tar.gz /etc
```

```
# Backup the voicemail directory
tar cfz /backup/asterisk-vm-${date}.tar.gz /var/spool/asterisk/
voicemail

# Rotate the logs for Asterisk
/usr/sbin/asterisk -rx 'logger rotate'

mv /var/log/asterisk/debug.0 /tmp/debug.${date}
mv /var/log/asterisk/messages.0 /tmp/messages.${date}
mv /var/log/asterisk/event_log.0 /tmp/event_log.${date}

# Backup log files
tar cfz /backup/asterisk-astlogs-$date.tar.gz /tmp/*.${date}

# Remove unnecessary files
rm -f /tmp/*.${date}

# Copy all archives to our backup server.
scp -B /backup/*-configs-${date}.tar.gz $backup_server
scp -B /backup/*-vm-${date}.tar.gz $backup_server
scp -B /backup/asterisk-astlogs-* $backup_server
scp -B /backup/asterisk-phonecfg-* $backup_server
```

Next, let's see `monitor_mix.cron`:

```
#!/bin/bash

####################################################
## Shell script to handle phone monitoring        ##
## files.  This script will be run at night       ##
## and maybe at lunch.  It will use soxmix        ##
## to mix the in and out components of the        ##
## conversation, delete the in and out com-       ##
## ponents, and then use lame to encode the       ##
## mixed wav file into an MP3                     ##
####################################################
## Usage: monitor_mix.cron, no arguments          ##
## required                                       ##
####################################################

# edit variable below to contain your
# backup server user and hostname as
# well as directory location
backup_server="username@backupserver:/path/to/backups"
date=`date +%Y-%m-%d`
```

```
# Clear previous backup files prepare folder
# structure for backup set.
rm /backup/monitor -rf
mkdir /backup/monitor
mkdir /backup/monitor/${date}
chmod -R 700 /backup/monitor

# For each conversation in the monitor directory
# soxmix the two parts of the conversation
# together and convert to mp3.
cd /var/spool/asterisk/monitor
for i in 'ls *-in.wav'
do
  basename='basename $i -in.wav'
  echo $basename

  # soxmix "in" and "out" files
  soxmix $i $basename-out.wav $basename.wav
  rm -f $i
  rm -f $basename-out.wav

  # convert resulting wav to mp3
  lame --resample 16 -m m -b 32 -h --cbr $basename.wav $basename.mp3

  # Remove unnecessary files
  rm -f $i
  rm -f $basename-out.wav
  rm -f $basename.wav

done

# Put newly created mp3s in local backup directory
mv *.mp3 /backup/monitor/${date}

# Copy mp3s to our backup server running an sshd
scp -B -r /backup/monitor/* $backup_server
```

We could also restart the Asterisk service every night in order to resolve any channels that may have hung. This may not be possible if the system sees high usage during the night, but is useful in situations where we have channels hanging often.

Finally, let's see `asterisk_restart.cron`:

```
#!/bin/bash
/etc/init.d/asterisk stop
/etc/init.d/dhadi restart
/etc/init.d/asterisk start
```

Time synchronization

As our Asterisk system retains logs of calls and also makes routing decisions based on time, it is important to have the system clock synchronized. We can do this with **Network Time Protocol (NTP)**. This is very easy to use. Just install the `ntpdate` program, which you will most likely find in your distribution's package management system (`yum`, `urpmi`, `apt`, or any other). Then run the script shown next. If we have a local time server (for instance, if we have other time-dependent services such as Kerberos authentication installed), we should use that.

The `timesync.cron` NTP script is as follows:

```
#!/bin/bash
ntpdate pool.ntp.org # replace pool.ntp.org with local time server
/sbin/hwclock --systohc # sync the hardware clock
```

Adding it all to cron

The four files we have created can be added to our crontab to ensure that they are run periodically. We can do this by first running:

```
$ crontab -e
```

Then creating the following entries (replacing `/path/to` with the location of our scripts):

```
30 01 * * *             /path/to/backup.cron
59 23 * * *             /path/to/monitor_mix.cron
00 01 * * *             /path/to/asterisk_restart.cron
00,15,30,45 * * * *     /path/to/timesync.cron
```

This ensures that our `backup.cron`, `monitor_mix.cron`, and `asterisk_restart.cron` scripts are run every night and that our time is synchronized every 15 minutes.

As we are running these commands non-interactively, that is we are using `scp` in batch mode, we must ensure that the `scp` command can authenticate with the backup server. In order to do this, we should use SSH keys instead of passwords for authentication by running the following commands:

```
$ ssh-keygen -t dsa ; accept all defaults by pressing enter at each
                    ; prompt
$ ssh username@backupserver 'cat >> ~/.ssh/authorized_keys'
   < ~/.ssh/id_dsa.pub ; enter password when prompted
```

Back up Asterisk with FreePBX

FreePBX has made the entire backup process much quicker and easier, all because of a module accessible through the GUI. By default, it might not be installed in FreePBX, so you will first need to install it through the **Module Admin** section. Once installed, you will find the **Backup & Restore** module located under the **Tools** section of the GUI.

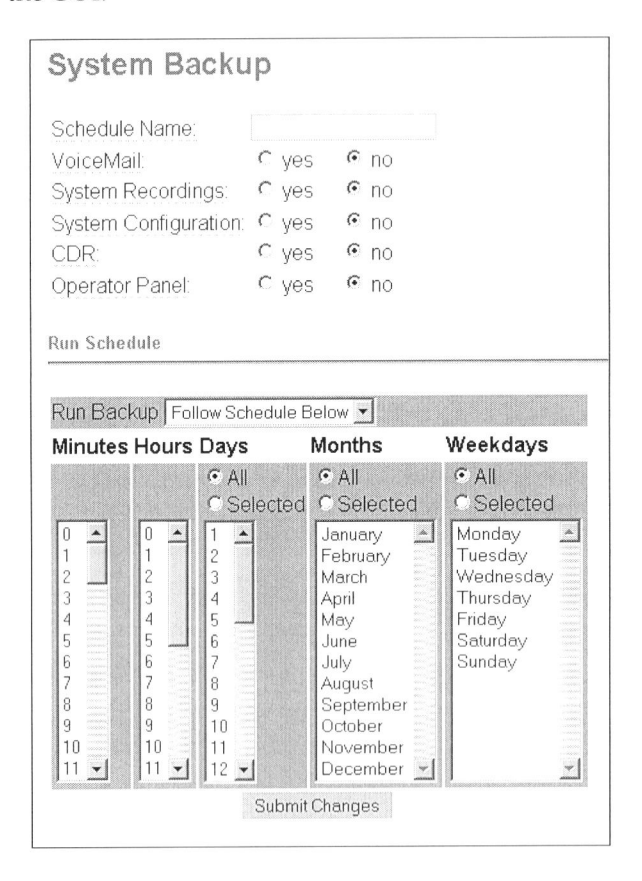

In the previous screenshot, you will see that you have the ability to create multiple backup operations quickly and easily. First, you will specify a name to easily identify the backup operation. Next, you need to specify what parts of Asterisk you wish to back up. For example, you might want to have a daily schedule that backs up system configuration, but for voicemails you might want to set up a separate schedule for once a week or once a month as each backup can be very large. The last part of the interface allows you to specify the days, months, and the time you want your backup to run. If you want to create an immediate backup, select "Now" from the drop-down menu labled **Run Backup**.

Back up Asterisk with Trixbox

Much like FreePBX, Trixbox also has a backup module, which is built into Trixbox and is accessible under the maintenance section (`<yourip>/maint` section). Much like FreePBX, the same settings are available to you in order to specify date and time as well as what parts of Asterisk you want backed up. The major difference is that Trixbox also has the option to specify a remote destination for the backup through FTP.

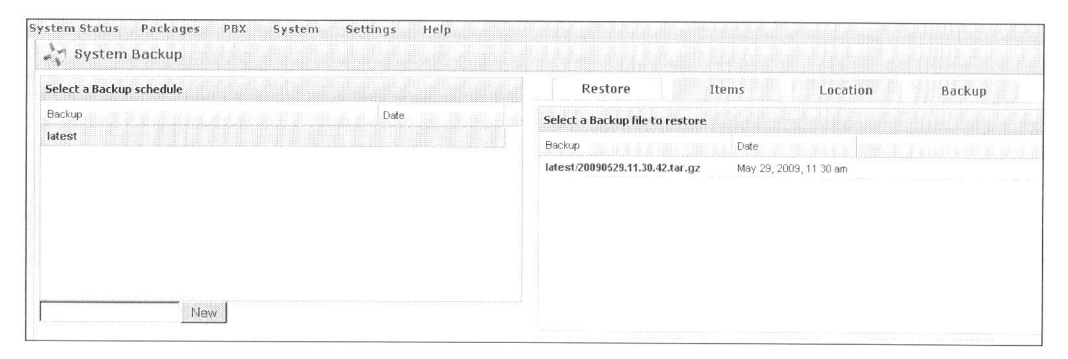

Rebuilding and restoring the Asterisk server

If the unthinkable happens and we lose our server due to hard disk failure or if we have to rebuild our server because of a system compromise, we need to know exactly how to get the server back online as fast as possible to minimize downtime.

There are a number of steps involved in this:

1. We rebuild the server, by following the instructions in the installation chapter. We follow the same process exactly up to the point of configuration. As we have a backup of the configuration files, we can skip this part and replace the files with our backups later.

2. We replace the configuration files. We identify the latest usable backup, from which we extract the `/etc/` directory. We then replace the operating system's configuration files we need and replace the `/etc/asterisk` directory. This ensures we have our previous configuration.

    ```
    #!/bin/bash
    $ tar xjvf asterisk_backup.tar.bz2
    $ cp -R etc/asterisk /etc/asterisk
    ```

    We follow a similar process for any other configuration files we wish to restore.

3. For data and logs, we follow the same process as in step 2, but this time we restore the `/var/spool/asterisk` and the `/var/log/asterisk` directories to their original locations as required, as well as any other areas of the system we have backed up.

4. Permissions are the last thing that we need to ensure so that Asterisk can read and write to the files necessary to function.

At this point, we are able to restart the Asterisk server and verify that the system works properly by testing whether we can make and receive calls and checking that all features of the system are functioning as they were previously. We could also ensure that no errors appear, after which it can be reintroduced back to its production environment.

Disaster Recovery Plan (DRP)

If during installation, we document as much as possible and create a valid DRP, we will be able to get our Asterisk server back online with minimal disruption and effort. As Asterisk is most likely an essential line of communication to partners and customers, downtime is an extremely important aspect to consider and creating a DRP should be addressed long before any disaster occurs.

Even something simple such as logging the installation process and documenting how to restore from backups is enough for at least a basic DRP, although it is recommended that we go further and create a full plan for not only the Asterisk server but also all other mission-critical services. We could also possibly have a layer of fault tolerance built into the system.

Our plan must take into account at least the following:

- Notification of any outage or data loss
- Times when outage of the Asterisk service would be most detrimental to the organization
- Responsibility for getting the service restored
- Estimated time scale for restoration of service
- Location of backups and other necessary files
- Vendor support contact details
- Detailed restoration instructions, which would include:
    - Restoring the service
    - Restoring all data
    - Restoring all configurations
    - Reimplementing backup procedures

Asterisk server security

Before we cover external security and think of putting the Asterisk server onto our production network, we must consider the internal security of the system to ensure that it fits with our security policy and meets good security practice at least.

Internal host security can be achieved in a variety of ways and there are many applications and tools that we can use to aid us in this. We will not discuss all of the tools and add-ons we can use for generic system security. However, we will cover basic operating system hardening with Asterisk in mind, as well as the further steps we can take to ensure that the Asterisk system is running as securely as possible.

It is also important to consider the physical security of the Asterisk system. We may want to have it under lock and key along with our other important infrastructure devices.

Internal access control

One of the most important and most overlooked aspects of host-level security is physical security. Our Asterisk server is a communication channel and most likely carries some confidential information. Be sure to have it as segmented from other non-essential and non-confidential systems as is reasonably possible.

In any multi-user system, internal **Discretionary Access Control Lists** (**DACLs**) are essential for security, and Linux as an OS has Unix foundations for these control lists. There are a variety of permissions that go far beyond read, write, and execute. However, focusing on these is enough for our purposes and will help us maintain a secure system. As a rule, we would have no one but administrators accessing the Asterisk server, because our users operate the system transparently from their telephony devices—either a handset or a software telephone. No direct access to the system is usually required. This is assuming that our Asterisk system is installed on a machine on its own that provides no other services, which is not always the case but is *highly* recommended.

Installing the Asterisk service on a dedicated machine offers the following benefits:

- Resources are dedicated to Asterisk and are therefore easier to monitor:
    - We know if Asterisk requires extra resources
    - Badly performing services don't affect Asterisk
- The attack surface of the machine is reduced:
    - There are fewer avenues for remote attack if the Asterisk service is the only way in

- Maintenance of the system is easier:
    - We don't have to check Asterisk updates for compatibility with other services and vice-versa
    - A system's reliability and uptime are inversely proportional to the number of services and users it provides for
    - There won't be downtime of the Asterisk service while unrelated services or components are updated, modified, or removed

We should ensure that the Asterisk service has access to the directories it requires to perform its function. The permissions mentioned here are not the default, but create a more secure setup, and we should test that our Asterisk service functions properly after making these changes. The default permissions on the directories listed are usually **755** and, for the files, **530**. The directory locations listed below are the default directories and permissions for an Asterisk install on Debian; consult your `asterisk.conf` file to confirm their exact locations on your distribution.

The key directories are:

1. `/etc/asterisk`: The configuration files for Asterisk are here. The Asterisk service requires read access to this directory so that it can read its configuration as it loads up and prepares itself for use. It won't need write access, as we will modify these files outside Asterisk and let Asterisk read them as it needs them.

2. `/usr/lib/asterisk/modules`: Asterisk has a variety of add-ons and different functionality provided by modules, which are shared libraries that Asterisk can load as needed. They are stored in this directory, and read access is all that's required. We don't require write or execute access as these modules aren't executed directly, but are loaded by an already running program (the Asterisk binary).

3. `/var/lib/asterisk`: This folder contains required files such as public keys for services. Read access is required so that Asterisk can read and present these keys to service providers. Write access is not required as when new keys are added or keys change, we would modify them manually.

4. `/usr/share/asterisk/`: This contains common files such as sounds. Again only read access is required so that Asterisk can load and play the sounds when necessary.

5. `/var/spool/asterisk`: This is the spool directory for storage of voicemail messages and other data. Here read/write access is required. Asterisk stores this data in real time and therefore, needs to constantly write to this directory. We wouldn't manually edit anything here usually, although we may occasionally delete old data.

6. `/var/run/asterisk`: Asterisk stores the PID of the currently running service here, and so it requires read/write access.

7. `/var/log/asterisk`: Asterisk keeps a variety of log files here and requires read/write access to continually log information and errors related to the Asterisk service.

   The following short script will set the permissions outlined above:

   ```
   #!/bin/bash
   # Sets minimum permissions required for Asterisk's key directories
   # Modify directory locations based on your /etc/asterisk.conf file

   # Make files root owned and asterisk grouped so that a compromise
   # of the asterisk user doesn't allow write access to the
   # config files

   chown -R root:asterisk /etc/asterisk /usr/lib/asterisk/
    /usr/share/asterisk/ var/spool/asterisk /var/run/asterisk
    /var/log/asterisk

   # Give owner (root) r/w/x and give group (asterisk) r/x

   chmod 750 /etc/asterisk /usr/lib/asterisk/ /usr/share/asterisk/
    /var/spool/asterisk /var/run/asterisk /var/log/asterisk


   #Additional write access for asterisk group on necessary
   #directories

   chmod 770 /var/spool/asterisk /var/run/asterisk /var/log/asterisk
   ```

If you have more Asterisk add-ons installed, the permissions of those files and directories will also be likely to require modification to increase security.

Host security hardening for Asterisk

When we have our basic DACLs in order, we can consider a number of other methods for keeping the Asterisk system secure.

There are several tools that can be installed and used to improve security on Asterisk, and describing the options for many of them would take up entire bookshelves of their own. Here, we will discuss some of the simpler tools for keeping you informed on how secure your system is.

Integrity checker

We could install Tripwire or another file integrity checker to monitor the checksums (hash values calculated from a file's contents) in order to ensure that the contents of a file haven't changed. This helps by informing us whenever a file changes; more specifically, it focuses on binary files. Hence, if an attacker succeeded in altering the Asterisk binary or one of the modules, you would know about it. You can also monitor other operating system files (`netstat`, `ps`, `top`, and so on) in order to ensure that they haven't been tampered with. The security offered in this is knowing which things have changed without your approval in the event of a system compromise.

> **Checksums**
>
> A checksum is calculated by running a file's binary contents through a known algorithm, giving a constant value as long as the file contents do not change (the filename has no relevance with the sum). This is used by tools such as Tripwire to determine if a file has changed.

Rootkit detection

Rootkits are tools installed by attackers in order to gain control of the system. They modify binary files, change kernel system calls, and use a variety of other evasion techniques such as covert communication channels. All this is in aid of keeping the attacker hidden so that we don't know they have compromised us, leaving them free to plunder the system and use our resources as they see fit. A rootkit detection tool is useful as it helps us find these rootkits and quite often helps remove them. The two most notable tools are **rkhunter** and **chkrootkit**.

> For more information on rkhunter, visit http://www.rootkit.nl/.
> For more information on chkrootkit, visit http://www.chkrootkit.org/.

Automated hardening

We can also use a tool such as **Bastille**, which will help us harden areas of the system outside of Asterisk. We can implement a host-based firewall and modify other system settings to increase security. Bastille has a wizard-based interface, which asks the user a series of questions, and then creates and applies a security policy based on the given answers. It requires very little knowledge of the underlying system and is a generous boost to the overall security of the Asterisk host.

There is a plethora of other tools to choose from; however, these are very common and very easy to use and are almost essential to a secure system. Installing and using these at a minimum provides the knowledge of what's going on within our system, which is an important part of knowing how secure we are.

Role Based Access Control (RBAC)

It's long been known that the traditional DACL, which is prevalent in many OSs including Windows and Linux BSD as well as other Unix-based systems is not the only way nor the best way to separate system access. RBAC is not an entirely new idea, and it has been around for a long time, but doesn't see much usage due to being quite difficult to implement.

Asterisk is a very complicated system, which performs a variety of functions, so it can be very difficult to create a workable access control list for access to system resources. RBAC works by not having a single root or administrative user, but instead splitting all tasks to only those users in the system that require them.

RBAC can be provided on Linux by using RSBAC (Rule Set Based Access Control) found at http://www.rsbac.org. You can use Adamantix, which is a distribution that has this already fully implemented, and there are configurations of the system available to set up Asterisk.

> For more information on Adamantix, visit http://www.adamantix.org.

> **SELinux**
> SELinux is a patch for the Linux kernel produced by the NSA for their purposes, and it is described as experimental. It is, however, used on a variety of production networks as an implementation of the Mandatory Access Control theory.

The good news for Asterisk administrators, however, is that there is a pre-written script for SELinux downloadable from http://www.coker.com.au/selinux/ if you use Debian. The policy is also available for other distributions such as Gentoo, where you will find it within Portage.

Network security for Asterisk

As many of the protocols Asterisk supports are used over a TCP/IP network, we need an understanding of how to control and firewall these correctly in order to ensure that we only let the necessary traffic pass through.

Our firewall will most likely be on a box separate from our Asterisk installation and placed at the network perimeter (we may also have a host-based firewall to which different rules may apply). In order to define the required rules, I won't detail how to configure a specific firewall product, but provide the details necessary to configure any device we have protecting our Asterisk installation.

These rules would apply to any device, be it `iptables` on a Linux machine, a commercial firewall such as Microsoft ISA server or checkpoint, PIX, and so on. The product in use isn't the main issue, the protocol rules that are required are. We can then take these generic rules and apply them to any firewall device we decide to install.

Firewalling the Asterisk protocols

When it comes to security, firewalls have traditionally been the most important mechanism for protecting internal company assets. For your Asterisk implementation and more specifically the VoIP elements of this system, this is an important consideration.

VoIP protocols are among the most complex in common use and require a great deal of forethought before we can go ahead and deploy Asterisk on our production network. We must consider which solutions we will use and which providers can supply them for us. In order to ensure the network is securely set up, we should have a thorough understanding of the protocols that we'll use so that we can firewall effectively.

It is often difficult to firewall VoIP protocols and there are many extensive documents detailing various scenarios, so here we will discuss the basic needs of a protocol for firewalling.

Probably the three most common protocols used by VoIP communications today are SIP, H.323, and IAX. The choice of protocols used for our VoIP communications depends entirely on which protocol our vendor supports and which protocol our contacts use.

We will cover SIP, H.323, and IAX here. We have covered these protocols from a technical perspective in previous chapters and we will now see how to firewall these effectively and why IAX is much easier to maintain from a network-control perspective.

Chapter 10

...rotocol)

...have SIP support built-in. All of the
...ough the device are available and all
...to be flicked. If we have such a firewall
...ASA products, and ISA server), then our
...fine our own firewall rules, a little more
...ditional firewall, which is a border control
...the LAN and the Internet), then it is relatively

...5060 (UDP and TCP) to the Asterisk machine

...5060 (UDP and TCP) from the Asterisk server

...on (NAT) between our Asterisk server and the
...tle more complicated. In order to get such a
...get a SIP Proxy that supports NAT, which will
...culty.

...is a hack that was created in order to increase
...NAT works by taking the internal address
...odifying packets sent out so that they use
...dresses. The NAT gateway then takes the
...l rewrites them for the original client based
...This works well for most single socket
applications (a socket being a pair of IP addresses and ports).

In SIP's case, this will not work as SIP requires a distinct address for the SIP client and when we use NAT this is obviously not the case, as multiple clients use the same address. The Internet Engineering Task Force (IETF, www.ietf.org) maintains the Internet drafts that detail exactly how to get SIP working through a NAT device. The best advice, however, is for us to use a SIP Proxy or attempt to route our SIP service through a router without NAT, that is, to basically give Asterisk one of our publicly accessible IPs. Depending on our placement of the Asterisk server this may be a viable solution. With IAX, we can link multiple Asterisk servers, so this gives us added options when it comes to server placement.

H.323

H.323 has a similar problem to SIP as it is also designed to require distinct IP addresses and the same advice applies—if we can have the H.323 server on a public IP then it may be easier to maintain it, as long as we firewall it effectively. If we have control of both ends of the communication, we can set up a VPN between the two sites, which solves this problem and ensures end-to-end encryption.

In order to firewall H.323, we need to permit incoming and outbound connections on TCP port 1720 and UDP ports 5000-5014.

IAX

IAX is a lot more straightforward than either H.323 or SIP as it was designed with the limits imposed by NAT in mind. You can easily allow this traffic through your firewalled NAT with minimal fuss.

IAX uses UDP port 4569 for outbound and inbound communication. The old IAX protocol, mentioned in an earlier chapter and succeeded by the current IAX (IAX2), used UDP port 5036.

IAX is also more powerful than either H.323 or SIP and has several features that make VoIP administration and use much easier. For example, it has enhanced signaling capabilities and separates signaling and data more effectively. Also, as IAX is not a standard and therefore has no standards body monitoring the decision process, modifications can be made more easily.

The Real-Time Transport Protocol (RTP)

RTP is the protocol often used to carry the audio data in a VoIP conversation. It is a standard developed by the IETF (Internet Engineering Task Force). It can also be used to carry video data and is designed specifically to handle this sort of real-time data. It attempts to guarantee that the data will be transmitted and received in a short period of time. Obviously latency in voice conversations can be a problem, so RTP avoids this latency as much as possible and concentrates on timely delivery of data.

In order to allow RTP to function, we would have to allow the UDP ports 10000 to 20000 inbound and outbound from our Asterisk server.

Controlling administration of Asterisk

As we have set up Asterisk to access files owned by the root user and Asterisk group, this means that the Asterisk service can read and write only to the files it requires. We, however, may have to perform additional maintenance tasks such as adding extensions, creating new voicemail boxes, and so on.

As Asterisk configuration is managed by modifying flat files, this is done by logging in to the server with an interactive session, at the local console or remotely. In order to follow best practice, we wouldn't log in directly as the root user, but more likely as the Asterisk user. If we did need to edit any files that the Asterisk user doesn't have privileges for, then we would switch user to root using the `su` command:

```
$ su -
Password:[root password]
```

We could also implement Sudo and give it access to our Asterisk user account. Either way would log in as root indirectly. In order to ensure that no one else can log in as root across the network, we should configure our remote access mechanism to disallow root logins. The most common remote access method for managing an asterisk server would be SSH and the most common implementation would be OpenSSH (for more information on OpenSSH, visit `http://www.openssh.com/`), which can be configured to prevent root login by editing the relevant directive in the configuration files:

```
$ cat >> "PermitRootLogin No" >> /etc/ssh/sshd_config
```

We could also secure our remote access further by using the internal firewall (set up earlier by Bastille) to allow access only from our administrative team's IP addresses. This would prevent external attackers and internal users from making unauthorized connections to the Asterisk server.

> **Sudo**
>
> Sudo allows us to give restricted administrator access to selected users, but be warned that it is quite easy to misconfigure and give away more access than you intend.
>
> For example, giving someone Sudo access to `vim` gives them the ability to write to all files as root and to execute a root shell from within `vim`. Most likely, this is not desirable!
>
> For information, visit `http://www.courtesan.com/sudo/`.

For those using FreePBX or Trixbox, the GUI is another piece you will want to protect. First, you have to decide if you wish to have your GUI accessible to the public Internet. Apache, which operates most open source GUI's for Asterisk, will operate over TCP port 80. It's best practice not to open this port to the outside world. However, if you must, then you might want to consider changing the port to something other than 80, such as 8085, among others. The reason for this is that hackers have applications that scan the Internet for those using Trixbox and other Asterisk GUI's, and once they detect a server they try to attack it to gain access. Their basic method is of course scanning on port 80. Hence, if you change your port to something other than 80 you're less likely to be detected from someone generally scanning IP addresses. Information on changing the port on Apache can be found in your `httpd.conf` file located in `/etc/httpd/conf/`.

Also keep in mind that after changing your port number you will need to access your GUI using that new port number within the URL, such as `http://<yourIP>:8085/admin`.

Asterisk scalability

As the Asterisk server is most likely highly critical to business, we want to ensure that restoring from backups rarely happens and in the event of losing a machine when an administrator isn't available, we have some sort of failover system in place. In order to achieve this, we apply redundancy and load-balancing techniques to ensure that our infrastructure has the resources to handle the data it needs to process.

In the event of a component failing, we would like to ensure that we don't lose services. Ideally, the users of the system should never know there was any failure and the administrator should get the failed system back online or replace it at the next convenient moment. This sort of forward planning is essential for maintaining a service that will be used as extensively as Asterisk often is.

Take the example of the 24-hour call center. If we have a business that relies on the telephony system in order to generate revenue, then the loss of that service is a loss of revenue. Being a 24-hour service, there may not always be an administrator on site—there may be periods where there is only "on call" cover. It would be a waste of resources to have all of the users idle while they wait on an administrator to possibly be wakened and then ferried to the site in order to get the system back up and running.

A single point of failure is not only undesirable but can also have a severe negative impact on the profitability of the business. There is also the chance that if our usage of the Asterisk system outgrows current resources, then the Asterisk machine has to be taken offline while it is upgraded. We could avert this by having scalability built into our design from the outset to ensure that the system can grow with business demand.

As Asterisk can't be installed onto a cluster, we require load balancing and scalability that can be implemented without the use of clusters, which isn't as hard as it might seem.

Load balancing with DNS

One of the most common ways to load-balance a system is to use DNS, the Domain Naming System. This has the ability to "round robin" replies to queries in order to spread load between different machines.

One of the largest DNS load balancing systems that we all have most likely used is the Google search engine.

```
$ dig google.com +short
216.239.37.99
216.239.57.99
216.239.39.99
$ dig google.com +short
216.239.57.99
216.239.39.99
216.239.37.99
$ dig google.com +short
216.239.39.99
216.239.37.99
216.239.57.99
```

As we can see, each time the command to look up the IP of the Google server is run, it returns a different IP than the first IP, so that clients accessing it are spread between all of the IPs in the pool and no single machine gets overloaded. You can also add addresses to the pool and remove them without affecting the client, which means this system will scale well to allow many users to access what appears to be just one service.

The advantage of using round-robin DNS is that the server hardware behind the service has no direct bearing on how the system can be scaled, as we have the option of adding and removing servers. For instance, if we were to suddenly grow, then we could add in more servers or replace/upgrade existing servers leading to very simple scalability. There is also some redundancy inherent in this system; if your client can't contact the first server, it will then attempt to contact the second server. This means that the loss of a server doesn't bring the entire system to a halt, it merely slows it down slightly.

> **Caching**
> As clients cache the IPs they get from DNS, when they find a working IP, the slow-down incurred will be negligible. However, it is highly recommended to remove problematic machines or addresses from the pool.

The example we look at here uses A records; however, it is increasingly common to see round-robin implementations for SRV records. SRV records are used to locate a service within a domain. For example, in Microsoft's Active Directory implementation, SRV records are used to locate domain controllers and we would use them to locate our routing service providers in our Asterisk setup. The functionality of round robin doesn't differ for the record you request. However, you still create a pool of IPs in your DNS implementation. The two most commonly used implementations of DNS, Microsoft DNS and BIND (versions above 4.9.7), support SRV records and round-robin SRV records.

For example, to set up multiple SRV records for our SIP implementation, we would add the following to our DNS zone:

```
sipa    IN    A         10.1.1.1
sipb    IN    A         10.1.1.2
sipc    IN    A         10.1.1.3

sip     IN CNAME        sipa.example.com
        IN CNAME        sipb.example.com
        IN CNAME        sipc.example.com

_sip._udp.example.com.    IN    SRV    20    0    5060    sip.example.com.
```

This sets up three SIP servers for us (`sipa`, `sipb`, and `sipc`) and one `_sip._udp` SRV record. Whenever the SRV record is requested, one of these three SIP servers will be returned.

Support channels for Asterisk

As an open source project, we would expect Asterisk to have at least some basic community support that we could rely on. Asterisk does have this and it has quite a bit more as well. It has mailing lists, forums, and IRC as well as official support from Digium. We don't always require commercial support. However, if running Asterisk is not our core responsibility or if we have other constraints, then having paid technical support on hand can be a resource we would welcome.

Mailing lists

There are a few mailing lists available for unofficial Asterisk support, by far the most active being provided by Digium itself. They are frequented by Digium staff as well as Asterisk users and are probably the best source of information when it comes to quick opinions or support from the community. They are found at http://lists.digium.com/mailman/listinfo.

The **USERS** mailing list is the best choice for support issues.

There is also the VOIPSEC mailing list provided by www.voipsa.org, which isn't Asterisk-centric as its main focus is VoIP security on a wider scale. However, as Asterisk is one of the most common VoIP solutions, it is a topic of frequent discussion on the list and topics such as firewalling protocols or encrypted communication are directly relevant to anyone responsible for the security of an Asterisk installation.

We may not decide to use these mailing lists as a support mechanism; however, it is worth having a "lurk" and reading through them at least, to give an insight into how other people are using Asterisk and the problems and issues they come across. Such experiences are invaluable in ensuring we do not repeat others' mistakes and will help in increasing our knowledge of Asterisk and associated technologies. The VOIPSEC list for instance has become the focal point of VoIP security and is often the first outlet for information that has an impact on the security of a VoIP implementation.

Forums

You can also obtain some support from the Digium forums, which can be found at http://forums.digium.com/. However, they aren't as busy as the other support available—the mailing lists and IRC being most popular.

Trixbox and FreePBX also have a very extensive forum containg a great deal of information on both the GUI as well as Asterisk; they even have sections for VoIP hardware.

Trixbox fourm can be found at http://www.trixbox.org/forum.

FreePBX forum can be found at http://www.freepbx.org/forums.

Maintenance and Security

Internet Relay Chat (IRC)

Asterisk has a lively community support mechanism provided by its IRC channel. This can be found on the Freenode network, which is a network that almost entirely comprises of support channels for free and open source software.

In order to access this, download a suitable IRC client. mIRC, X-Chat, irssi, and chatzilla are commonly used clients, and most have the address for the Freenode (`irc.freenode.net`) servers in their default configuration. Once connected to Freenode, join `#asterisk`. This channel is much like the Digium mailing list, in that it focuses on the discussion of the use and administration of Asterisk. It is also frequented by the same people that use the mailing lists—developers, administrators, and users.

Web sites

You can also obtain a great deal of information on Asterisk as well as other SIP applications and hardware from Voip-Info (`http://www.voip-info.org`). They have a very impressive collection of information on hardware devices, VoIP service providers, as well as information on both commercial and open source applications for Asterisk.

For those looking for easy-to-follow video tutorials and step-by-step instructions on configuring devices for Asterisk and using GUI's such as Trixbox and FreePBX, you will want to check out ThinkBright, which is a VoIP business service provider, but also hosts a free video tutorial section. They can be found at `http://www.thinkbright.net/support/`.

IRC becomes a valid support mechanism when we need quick short answers to our problems or a quick sanity check on something we intend to do. It is often difficult to solve complex problems on IRC, especially those that require long detailed explanations or which extend over large portions of our configuration files. It's often easier to explain such matters in an email to a list. However, IRC is good for quick replies and for question and answer sessions, so it shouldn't be overlooked.

Digium

Digium provides a variety of services relating to the installation and running of Asterisk, from email and telephone support to on-site support contracts. We should evaluate the need and benefit of this as we decide which kind of support we need. For example, if we have a full-time Asterisk administrator or team, we probably wouldn't require much support; maybe email support for occasional troubles, but the mailing lists could possibly provide enough.

If, however, we are employed as a single administrator in an SME, we would benefit from having official support mechanisms on hand, although in reality the spread of support is usually the other way around, with SMEs winging it and larger companies having too much support. We also have to consider cost; unofficial community support will obviously be cheaper than paid commercial support. We should evaluate our needs carefully and ensure that we have the necessary support in place to maintain our Asterisk system.

Summary

The phone system of any modern business is something that, if it works well, should be almost invisible to its users. We want them to take it for granted, and to use its features without thinking. It's inevitable and even desirable that our users should come to depend on the services the system offers. Naturally therefore, we want to minimize any disruption to the system, and to make sure that, in the event of a failure, normal service can be resumed as smoothly and quickly as possible.

In this chapter, we've looked at how to be prepared for such an eventuality, by performing regular and systematic backups. We also looked at making a Disaster Recovery Plan, which can help to minimize the time taken to get the system back online.

Of course, the best way to minimize disruption from service outages is to prevent them from happening in the first place. To this end, we have looked at how to make Asterisk more robust and how to harden it against attack.

Not all failures are the result of malicious activity, however, and we've also covered a few issues that you should consider in order to make Asterisk scale well. Finally, the community support channels are invaluable in keeping your Asterisk system well maintained and running efficiently, as well as providing help should you ever get stuck. The last section of this chapter was devoted to coverage of these various channels.

Index

Symbols

\# key 100
- b option 116
-c command line argument 54
- e(ext) option 117
- g(grp) option 116
-o option 117
-q option 117
- r[(basename)] option 117
- v([value]) option 117
-v command line argument 54
-w option 117
-x option 117
<device>, options
  bchan 62
  dchan 62
  e&m 62
  fxsgs 62
  fxsls 62
  unused 62
rsyncd.conf file 189

A

Advanced Call Distribution
  about 96
  call parking 100
  call queues 96
  DID 101
Analog Terminal Adapter. *See* ATA
announce-frequency variable 83
announce-holdtime variable 83
announce variable 81
ARI 137
asterCC
  about 149
  downloading 149
  installing 150, 151
  manual installation 151-157
  prerequisites 150
asterCRM
  about 149
  customer section 159
  dialer section 159
  extension section 158
  import section 157
  interface 150
  statistic section 158
  survey section 160
  system section 159
asterCRM installation
  automatic installation 151
  manual installation 151
Asterisk. *See also* Asterisk 1.6, features
Asterisk
  about 7
  Appliance 16
  Asterisk 1.4, comparing 8
  Asterisk 1.6, comparing 9
  Asterisk Mail 77
  backup 187
  basic behaviors 54
  call centre system 13
  cdr_csv module 112
  Comedian Mail, voicemail program 77
  conference rooms 83
  configuration files 50
  configuring 57, 58
  considerations 19
  deployment, planning 25
  downloading, commands 48
  FreePBX 18

hardware requirements 42
IAX interfaces 74
installing 47, 50
IVR system 13
MP3, streaming files to handset 80
music on hold 80
PBX 9
PBX, differentiating 11
preparing, to install 47
protocols, supported 33
queues 81
reload 55
restart 55
restarting, options 55, 56
sample configuration files 51
scalability 18
security 199
server security 199
SIP interfaces 70
starting 54
starting with 54
support channels 210
system maintenance 187
Trixbox CE 17
unable, to run on Windows 19
voicemail 77
voicemail system 14
VoIP system 14-16
Asterisk, considerations
ROI 22
TCO, calculating 21
trade-offs 19
asterisk-addons distribution 114
asterisk_restart.cron script 194
Asterisk 1.4, features
built-in voicemail system 8
generic jitter buffer 8
improved sound prompts 8
ITU standard T.38 fax calls, passing through 8
Jabber and Google Talk, IM support 8
shared line appearance 8
whisper paging 8
Asterisk 1.6, features
asynchronous events support 9
improved NAT support 9
improved reporting 9

new bridge 9
STUN support 9
Asterisk as PBX
Advanced Call Distribution (ACD) 11
call, recording 12
call barging 13
call parking 12
CDR 11
extension-to-extension calls 9, 10
line trunking 10
telco features 11
Asterisk backup
with FreePBX 196
with Trixbox 197
Asterisk deployment
hardware requirements 42
planning 25
terminal devices, choosing 38-41
terminal devices, types 31
Asterisk installation
adsi.conf file 51
agents.conf file 51
alarmreceiver.conf file 51
alsa.conf file 51
asterisk.adsi file 51
asterisk.conf file 51
cdr_manager.conf file 51
cdr_odbc.conf file 51
cdr_pgsql.conf file 51
cdr_tds.conf file 51
chan_dahdi.conf file 53
DAHDI, installing 49
enum.conf file 51
extconfig.conf file 51
extensions.conf file 51
features.conf file 51
festival.conf file 52
iax.conf file 52
iaxprov.conf file 52
indications.conf file 52
LibPRI 50
LibPRI, installing 49
logger.conf file 52
manager.conf file 52
meetme.conf file 52
mgcp.conf file 52
modem.conf file 52

modules.conf file 52
musiconhold.conf file 52
osp.conf file 52
oss.conf file 52
phone.conf file 52
prereqisites packages 47
prerequisite packages 47
privacy.conf file 53
procedure 50
queues.conf file 53
res_config_odbc.conf file 53
res_odbc.conf file 53
rpt.conf file 53
rtp.conf file 53
sip.conf file 53
skinny.conf file 53
source files 48
source files, obtaining 48
steps 47-50
telecordia-1.adsi file 53
voicemail.conf file 53
vpb.conf file 53
Zaptel 49
Asterisk protocols, firewalling
 H.323 206
 IAX 206
 RTP 206
 SIP 205
Asterisk Recording Interface. *See* **ARI**
Asterisk scalability
 about 208
 load balancing, with DNS 209, 210
Asterisk server
 rebuilding 197, 198
 restoring 197, 198
Asterisk server security
 about 199
 asterisk.conf file 200
 DACLs 199
 directories 200
 host security 201
 internal access control 199
 network security 204
 permissions, script 201
Asterisk service
 installing 199, 200
AsteriskWin3 19

ATA 29
automated attendants 103-106

B

backup.cron script 192, 193
backup and system maintenance, Asterisk
 backing up with FreePBX 196
 backing up with Trixbox 197
 backup schedule 189
 backup scripts 192
 configuration backup 188, 190
 DRP 198
 log files backup 191
 package management system 188
 server, rebuilding 197
 server, restoring 198
 timesync.cron NTP script 195
 time synchronization 195
 voice data backup 191
 voice data backup, rsync used 191
 voice data backup, tar used 191
backup scripts 192
 asterisk_restart.cron script 194
 backup.cron script 192
 monitor_mix.cron script 193
Bastille 202
bchan, <device> option 62
BRI, basic rate interface 27
Bring Your Own Device. *See* **BYOD**
BYOD 29

C

call centre system, Asterisk 13
Call Details Recording. *See* **CDR**
call parking
 about 12, 100, 101
 example 12
call queues 96-100
calls
 monitoring 116
 recording 118
calls, monitoring 116
 ChanSpy, executing options 116, 117
 ChanSpy, using 116
 DAHDIBarge, using 116
calls, recording 118

accessing 118
queues.conf file, using 118
CAS 27
case study
  hosted PBX 178
  small business 167
  small office/home office 161
case study, small business
  configuration, extensions 174
CDR
  about 9, 11, 112
  CSV file, benefit 112
  CSV file, order 112, 113
  database logging 113-115
  features 112
  flat-file logging 112, 113
  uses 12
CDR 112
  cdr-csv module 112
  flat-file CDR logging 112
  security scenario 112
cdr_csv module 112
cdr_pgsql.conf 114
CentOS 122
chan_dahdi.conf, DAHDI interfaces
  about 63
  accountcode option 67
  amaflags option 67
  busycount option 67
  busydetect option 67
  cadence option 67
  callerid option 67
  callgroup option 66
  callprogress option 67
  callreturn option 66
  callwaitingcallerid option 66
  callwaiting option 66
  cancallforward option 66
  channel option 67
  context option 64
  debounce option 65
  echocancel option 66
  echocancelwhenbridged option 66
  echotraining option 66
  flash option 65
  group option 66
  hidecallerid option 66

idledial option 67
idleext option 67
immediate option 67
jitterbuffers option 67
language option 64
lines 68
lines, grouping 68
mailbox option 66
minidle option 67
minunused option 67
musiconhold option 67
overlapdial option 65
pickupgroup option 66
preflash option 65
prewink option 65
pridialplan option 64
progzone option 67
relaxdtmf option 66
restrictcid option 66
rxflash option 65
rxwink option 65
signalling option 65
start option 65
switchtype option 64
terminals 68, 69
threewaycalling option 66
transfer option 66
usecallerid option 65
usecallingpres option 66
 usedistinctiveringdetection option 65
wink option 65
Channel Associated Signaling. *See* **CAS**
channels 62
checksum 202
chkrootkit 202
Comedian Mail, voicemail program
  about 77
  configuring 77
  format 77
  message length, limiting 78
  time zone messages, defining 78
  voicemail.conf 77
  voicemail box, example 79
communication devices, terminal
   equipment 36
conference rooms
  configuring 84

conf files
  about 54
  sample files 51
configuration files, samples 51
connection methods, PSTN
  E1 27
  ISDN 26
  POTS line 26
  T1 27
  VoIP connection 28, 29
context
  about 85
  creating 86, 87
context variable 82
CRM 149
Customer Relationship Management System. *See* **CRM**

D

DACLs 199
DAHDI
  installing 49
DAHDI interfaces
  chan_dahdi.conf 63
  system.conf 58
database CDR logging 113
dchan, <device> option 62
device, terminal equipment
  Asterisk compatibility 40
  choosing 38-41
  cost determination, handset cost 39
  cost determination, headset cost 40
  cost determination, installation cost 40
  cost determination, port cost 40
  cost determination, software license cost 40
  phone testing, feature 39
  requirement documentation, feature 39
  requirement elicitation, feature 38
  requirement prioritization, feature 38, 39
  sound quality, analyzing 40, 41
  usability issue 41
dialplan
  Advanced Call Distribution 96
  automated attendants 103
  context, creating 85, 86
  creating 85
  extension, creating 87, 88
  outgoing extensions, creating 92-94
  system services 106
DID 101, 102
DID numbers 30
Digium 212
Direct Inward Dialing. *See* **DID**;
Discretionary Access Control Lists. *See* **DACLs**
DRP 198

E

e&m, <device> option 62
extension
  common actions 88, 89
  creating 87-91
  fax calls 88
  invalid extension 88
  start extension 88
  timeout extension 88
extension length
  expected numbers 45
  1-digit, choosing 43
  another extension, adding 44
  choosing 43
  example 44
  expected numbers 45

F

files mode, music on hold 80
Flash Operator Panel. *See* **FOP**
flat-file CDR logging 112
FOP
  about 137
  features 138
  functions 138
Foreign eXchange Office. *See* **FXO**
forums 211
FreePBX
  about 121, 124
  downloading 121
  flexibility 143
  FreePBX System Status Dashboard 131
  installing 125
  prerequisites 125
  Web MeetMe 139

FreePBX installation
  about 125
  ARI password, fixing 129
  Asterisk, installing 127
  CentOS, installing 125
  codecs, adding to config 129
  extra packages, installing 126
  FreePBX, installing 127, 128
  MOH directory, fixing 131
  network settings, editing 125, 126
  Selinux, disabling 126
  Sendmail, configuring 129
  sip_nat.conf, editing 129
  voicemail config, editing 130
FreePBX System Status Dashboard
  about 131
  ARI 137
  Flash operator configuration files 139
  FOP 137
  getting 132
  reports section 136
  setup section 133
  tools section 132
  Trixbox maintenance section 135
FXO 26
FXO signaling 63
fxsgs, <device> option 62
fxsls, <device> option 62

H

H.323 33, 206
hard phone, terminal equipment
  Aastra 57 Series 32
  Cisco IP Phones (7940 & 7960) 32
  Grandstream GXP Series 31
  H.323 protocol 33
  IAX protocol 34
  Linksys SPA Series 32
  SIP 34
hardware phone. *See* **hard phone, terminal equipment**
hardware requirements
  CPU 42
  Pentium 90 42
  RAID-1 controller 42
  redundant server 43
  stable chipset 42
  UPS 42
hosted PBX, case study
  chan_dahdi.conf file 179
  conclusions 178, 185
  configuration 179
  extensions.conf file 182-185
  musiconhold.conf file 180
  planning 179
  sip.conf file 180, 181
  system.conf file 179
  voicemail.conf file 181
host security, Asterisk
  file integrity checker 202
  integrity checker 202
  RBAC 203
  rootkit detection 202
  system areas, harderning 202
  Tripwire 202

I

IAX 206
iax.conf 75
IAX interfaces
  about 74
  accountcode option 75
  amaflags option 75
  bandwidth option 75
  bindaddr option 75
  global options 75
  jitterbuffer option 75
  language option 75
  port option 75
  register option 75
  tos option 75
  trunkfreq option 75
  user entry options, setting 76
IAX protocol
  about 34
  advantages 35
  global options 75
  interfaces 74
  users, defining 76
installing
  asterCC 150
  DAHDI 49
  LibPRI 49

Integrated Service Digital Network. *See* ISDN
Inter-Asterisk eXchange. *See* IAX protocol
Interactive Voice Response. *See* IVR system
Internet Relay Chat. *See* IRC
IRC 212
ISDN
  about 26, 27
  BRI 27
  channels 26
  PRI 27
IVR system 13

J

jitterbuffer option, IAX interfaces 75

K

Key Systems 9

L

legal concerns 119
LibPRI
  about 50
  installing 49
Local Exchange Carrier (LEC) 26

M

mailing lists
  about 211
  USERS mailing list 211
  VOIPSEC mailing list 211
maxlen variable 83
meetme.conf 83
member variable 83
Message Waiting Indicator. *See* MWI
monitor_mix.cron script 193, 194
MP3, streaming files to handset 80
music on hold
  about 80, 81
  directory directive 80
  Mode directive 80
musiconhold.conf 80
music variable 81
MWI 8

N

Network Address Translation (NAT) 34
network security, Asterisk
  about 204
  Asterisk protocols, firewalling 204
  configuration management 207, 208
ntpdate program 195

O

outgoing extensions
  creating 92-96

P

PBX
  about 9
  communication devices 36
  hard phone 31
  soft phone 35
Plain Old Telephone Service. *See* POTS line
postgres_cdr.sql script 114
PostgreSQL 114
POTS line
  about 17, 26
  FXO, requiring 26
POTS line, connection method 26
PRI 27, 50
Primary Rate ISDN. *See* PRI
Private Branch Exchange. *See* PBX
PSTN
  about 10, 25
  E1 connection method 27
  ISDN 26
  needs, determining 29, 30
  POTS line 26
  T1 connection method 27
  VoIP connection 28
Public Switched Telephone Network. *See* PSTN

Q

QoS 33
Quantity of Service. *See* QoS
queues

about 81
members, defining 83
music, setting 81
queues.conf 81
ringall strategy, using 82
ring types, using 82
variables 81
queues.conf 81

R

RBAC 203
Real-Time Transport Protocol. *See* RTP
reload 55
reports section, FreePBX System Status
 Dashboard 136
restart, Asterisk
 about 56
 options 55
retry variable, queues.conf 82
Return on Investment. *See* ROI
rkhunte 202
ROI 22
Role Based Access Control. *See* RBAC
rootkit 202
RTP 206

S

sample configuration files 51
scalability 18
SELinux 203
Session Initiation Protocol. *See* SIP;
setup section, FreePBX System Status
 Dashboard
 extensions 133
 general settings 134
 inbound routes 133
 IVR 134
 on hold music/system recordings 134
 outbound routes 134
 queues 134
 ring groups 134
 time conditions 135
 trunk 134
simple PBX
 creating 143
 extensions, configuring 144

incoming route, creating 146
outgoing route, creating 147
routes 146, 147
trunks, configuring 145
SIP
 about 34, 205
 advantage 34
 configuring 70
 global options 70
 interfaces 70
 users, defining 72
sip.conf 70
SIP interfaces
 about 70
 accountcode option 71
 amaflags option 71
 bindaddr option 70
 call-limit option, defining 73
 callerid option, defining 72
 callgroup option, defining 73
 codecs option 72
 configuring 70
 context option 70
 defaultexpiry option 71
 defaultip option, defining 73
 deny option, defining 73
 dtmfmode option, defining 73
 externip option 71
 fromuser option, defining 72
 global options, configuring 70, 71
 host option, defining 72
 language option 71
 localnet option 72
 mailbox option, defining 73
 maxexpiry option 71
 md5secre option, defining 72
 musicclass option 71
 nat option, defining 73
 notifymimetype option 71
 pedantic option 71
 permit option, defining 73
 pickupgroup option, defining 73
 port option 70
 qualify option, defining 73
 realm option 70
 relaxdtmf option 71
 rtpholdtimeout option 71

rtptimeout option 71
secret option, defining 72
srvlookup option 70
type option, defining 72
username option, defining 72
videosupport option 71
small business, case study
  about 167
  agents.conf file 169
  chan_dahdi.conf file 169
  conclusions 178
  configuration 168
  extensions.conf file 173-177
  meetme.conf file 172
  musiconhold.conf file 169
  planning 167
  queues.conf file 170
  scenario 167
  sip.conf file 171, 172
  system.conf file 168
  voicemail.conf file 172
small office/home office, case study
  chan_dahdi.conf file 163, 164
  conclusions 166
  configuration 162
  extensions.conf file 165, 166
  modules.conf file 165
  musiconhold.conf file 164
  planning 162
  scenario 161
  system.conf file 163
  voicemail.conf file 164
soft phone, terminal equipment
  about 35, 36
  advantage 36
source files
  downloading, commands 48
  obtaining 48
span
  coding 61
  framing 61
  Line Build Out (LBO) 60
  number 60
  timings 60
spans 60
strategy variable 82
su command 207
sudo, Asterisk access restriction 207

support channels, Asterisk
  about 210
  Digium 212
  forums 211
  IRC 212
  mailing lists 211
  web sites 212
system.conf, DAHDI interfaces
  configuring 59
  line, channels 62
  lines 59
  signaling, identification 60
  terminals 63
system maintenance, Asterisk
  about 187
  areas 187
system services
  about 106
  conference rooms 108
  voicemail 107

T

T1, defining as span 60
TCO
  about 21
  annual licensing cost 22
  annual support cost 22
  installation cost 21
  licensing cost 21
  procurement cost 21
telephone systems 111
terminal equipment
  about 31
  analog adapter 36
  analog adapter, disadvantage 37
  analog adapter, use 37
  another PBX 37
  device, choosing 38
  selecting, to use 41
  types 31
  types, hard phone 31
  types, soft phone 35
test environment
  preparing 46
timeout variable, queues.conf 82
timesync.cron NTP script 195

time synchronization
    about 195
    fles, adding to crontab 195
tools section, FreePBX System Status Dashboard
    about 132
    Asterisk API 133
    Asterisk CLI 133
    Asterisk info 133
    Asterisk log files 133
    backup & restore 133
    print extensions 133
    system status 132
Total Cost Of Ownership. See **TCO**
trade-offs
    about 19
    flexibility, versus usage ease 19, 20
    graphical, versus file management configuration 20
Tripwire 202
Trixbox
    about 122
    features 122
    installing 123, 124
    prerequisites 122
Trixbox maintenance section
    backup 135
    bulk extensions 136
    config edit 135
    endpoint manager 135
    packages 135
    phpMyAdmin 135
    sysinfo 135

U

Uninterrupted Power Supply (UPS) 42
unused, <device> option 62
USERS mailing list
    mailing lists 211

V

voicemail
    attach option 79
    global options, configuring 77, 78
    maxlogins 78
    operator option 79
    saycid option 79
    silencethreshold 78
    skipms 78
    tz option 79
    zone messages, example 79
Voice over Internet Protocol. See **VoIP system, Asterisk**
Voice Service Providers. See **VSPs**
VOIPSEC mailing list
    mailing lists 211
VoIP system, Asterisk
    about 14-16
    cost, slashing 15
    cost based calls, routing 16
    networks, merging 14
    phone linking, benefit 16
VSPs 29

W

Web MeetMe
    about 139, 140
    accessing, through Trixbox 141, 142
    setting up 141
web sites 212

X

X-Chat 212
X-Lite 119

Z

zapata.conf. See **Zaptel**
Zaptel
    channels, configuring 62, 68
    configuring 58
    global options 59
    installing 49
    interfaces 58
    lines, device class 59
    T1, <device> options 62
    T1, defining as span 60
    terminals, device class 63
    zapata.conf, lines 68
    zapata.conf, options 64
    zaptel.conf 58
zaptel.conf. See **Zaptel**

Thank you for buying Asterisk 1.6

Packt Open Source Project Royalties

When we sell a book written on an Open Source project, we pay a royalty directly to that project. Therefore by purchasing Asterisk 1.6, Packt will have given some of the money received to the Asterisk project.

In the long term, we see ourselves and you—customers and readers of our books—as part of the Open Source ecosystem, providing sustainable revenue for the projects we publish on. Our aim at Packt is to establish publishing royalties as an essential part of the service and support a business model that sustains Open Source.

If you're working with an Open Source project that you would like us to publish on, and subsequently pay royalties to, please get in touch with us.

Writing for Packt

We welcome all inquiries from people who are interested in authoring. Book proposals should be sent to author@packtpub.com. If your book idea is still at an early stage and you would like to discuss it first before writing a formal book proposal, contact us; one of our commissioning editors will get in touch with you.

We're not just looking for published authors; if you have strong technical skills but no writing experience, our experienced editors can help you develop a writing career, or simply get some additional reward for your expertise.

About Packt Publishing

Packt, pronounced 'packed', published its first book "Mastering phpMyAdmin for Effective MySQL Management" in April 2004 and subsequently continued to specialize in publishing highly focused books on specific technologies and solutions.

Our books and publications share the experiences of your fellow IT professionals in adapting and customizing today's systems, applications, and frameworks. Our solution-based books give you the knowledge and power to customize the software and technologies you're using to get the job done. Packt books are more specific and less general than the IT books you have seen in the past. Our unique business model allows us to bring you more focused information, giving you more of what you need to know, and less of what you don't.

Packt is a modern, yet unique publishing company, which focuses on producing quality, cutting-edge books for communities of developers, administrators, and newbies alike. For more information, please visit our website: www.PacktPub.com.

FreePBX 2.5 Powerful Telephony Solutions

ISBN: 978-1-847194-72-5 Paperback: 292 pages

Configure, deploy, and maintain an enterprise-class VoIP PBX

1. Fully configure an Asterisk PBX without editing the individual text-based configuration files
2. Add enterprise-class features such as voicemail, least-cost routing, and digital receptionists to your system
3. Secure your PBX against intrusion by managing MySQL passwords, FreePBX administrative accounts, account permissions, and unauthenticated calls

Asterisk Gateway Interface 1.4 and 1.6 Programming

ISBN: 978-1-847194-46-6 Paperback: 220 pages

Design and develop Asterisk-based VoIP telephony platforms and services using PHP and PHPAGI

1. Develop voice-enabled applications utilizing the collective power of Asterisk, PHP, and the PHPAGI class library
2. Learn basic elements of a FastAGI server utilizing PHP and PHPAGI
3. Develop new Voice 2.0 mash ups using the Asterisk Manager
4. Add Asterisk application development skills to your development arsenal, enriching your market offering and experience

Please check **www.PacktPub.com** for information on our titles

trixbox CE 2.6

ISBN: 978-1-847192-99-8   Paperback: 344 pages

Implementing, managing, and maintaining an Asterisk-based telephony system

1. Install and configure a complete VoIP and telephonic system of your own; even if this is your first time using trixbox
2. In-depth troubleshooting and maintenance
3. Packed with real-world examples and case studies along with useful screenshots and diagrams
4. Best practices and expert tips straight from the Community Director of trixbox, Kerry Garrison

AsteriskNOW

ISBN: 978-1-847192-88-2   Paperback: 204 pages

A practical guide for deploying and managing an Asterisk-based telephony system using the AsteriskNOW Beta 6 software appliance

1. Install an Asterisk-based telephony system fast
2. Build an office PBX using AsteriskNOW
3. Learn the AsteriskGUI web management interface
4. Configure IP phones and connections
5. Configure and use the conferencing system
6. Write your own applications for Asterisk

Please check **www.PacktPub.com** for information on our titles

1124527R0

Printed in Great Britain by
Amazon.co.uk, Ltd.,
Marston Gate.